Michael Brooks, author of the acclaimed *13 Things That Don't Make Sense*, holds a PhD in quantum physics. He is a journalist and broadcaster, and acts as physics and cosmology consultant to *New Scientist* magazine. He has lectured at Cambridge University, the American Museum of Natural History and New York University, and his writing has appeared in many national newspapers, including the *Guardian*, *Independent*, *Observer* and *Times Higher Education*. He lives in Lewes, East Sussex.

The Big Questions confronts the fundamental problems of science and philosophy that have perplexed enquiring minds throughout history, and provides and explains the answers of our greatest thinkers. This ambitious series is a unique, accessible and concise distillation of humanity's best ideas.

Series editor **Simon Blackburn** is Distinguished Research Professor at the University of North Carolina, and Professor of Philosophy at the New College of the Humanities. He was, for many years, Professor of Philosophy at the University of Cambridge, and is one of the most distinguished philosophers writing today. He is the author of the bestselling books *The Oxford Dictionary of Philosophy, Think, Being Good, Lust, Truth: A Guide for the perplexed* and *How to Read Hume*. He lives in Cambridge.

Can We Travel Through Time?

The 20 Big Questions of Physics

MICHAEL BROOKS

SERIES EDITOR
Simon Blackburn

Quercus

This paperback edition published in 2012 by
Quercus
55 Baker Street
Seventh Floor, South Block
London
W1U 8EW

Hardback edition originally published by Quercus 2010 as
THE BIG QUESTIONS: Physics

A CIP catalogue record for this book is available
from the British Library

ISBN 978 1 78087 589 7

10 9 8 7 6 5 4 3 2 1

Text and plates designed and typeset by Ellipsis Digital Ltd

Printed and bound in Great Britain by Clays Ltd. St Ives plc

Contents

Introduction

The beauty of physics is summed up in one simple fact: a child can ask questions that no professor can answer. Indeed, searching out the 'big questions' in physics is rather like looking for hay in a haystack. When it comes to physics, it appears there is no such thing as a small question. A seemingly insignificant query or experiment can often lead to profound insight.

It is a short step, for instance, from asking whether the laws of physics can ever change or be broken to wondering whether there is room for a creator. It doesn't stop there, either. Physics tells us a creator need not be divine; it could be that we live nested within an infinite number of universes, each created by a species only slightly more intelligent than its greatest creation. We may even be destined to become creators of a universe ourselves.

With such big issues at its fingertips, it is small wonder that the most iconic scientists of our generation have been immersed in physics. Albert Einstein became a celebrity almost overnight when his theory of relativity changed our conception of the universe. Carl Sagan's TV programme *Cosmos* remains the most-watched series on public television. Richard Feynman's cool appraisal of the physics behind the *Challenger* shuttle disaster revealed how powerful a working knowledge of the subject can be. Stephen Hawking's work, laid out in his bestseller *A Brief History of Time*, created a thirst for scientific insight in people who had never given the sciences a thought. Only the discoverers of DNA, perhaps, can stand alongside these giants.

And yet, it has to be said, people also tend to recoil from physics. If I mention in casual conversation that I am a physicist by training, the announcement is met with a strange mixture of admiration and embarrassment. While expressing awe at anyone who would attempt to understand the universe, many also seem to consider the subject completely beyond them. 'Oh,' they say, 'I never did understand physics.'

If you recognize yourself in that statement, then hopefully this book will change your perspective. Perhaps the best-kept secret in physics is that there is too much there for anyone to understand. This is not a problem, however: this is the root of its allure.

Physics has so much to explore that, once it captures your imagination, it is hard to tear yourself away. The clock on the wall becomes a tease about the elusive nature of time. Sunshine is what results from a beautiful, intricate dance of particles known as nuclear fusion. When raindrops fall to the ground, you can ask yourself a simple 'why?' Exploring the answer will keep you occupied through the longest thunderstorm. The way a sunflower grows speaks of the conservation of energy and how the nature of light has shaped life on Earth. Go a step further and ask what light is, and you are peering into something widely considered to be the deepest mystery in nature.

This book is designed to show how simple questions lead to some of the most profound discoveries that humanity has ever made. They encompass the physics you probably didn't learn in class: the real point of the subject; its implications; what we understand about the universe – and what we don't. Carl Sagan once said, 'Somewhere, something incredible is waiting to be known.' Hopefully, that process can begin here.

What is the point of physics?

Impossible questions, unexpected rewards, and the never-ending quest for understanding

The question has bounced around school classrooms for decades. The answer offered usually starts with an apocryphal tale involving the legendary Greek philosopher Archimedes and King Hiero's crown.

Hiero had come to the throne in the Sicilian city of Syracuse. He gave a craftsman a certain quantity of gold to fashion into a crown; when the crown arrived, so did a rumour that the craftsman had substituted some of the gold for silver. Hiero commissioned Archimedes, then in his early twenties, to find the truth.

The story, as related by the Roman writer Marcus Vitruvius Pollio, says that Archimedes realized how to solve the problem when he noticed the volume of water that his body displaced in a bath. Silver, being less dense than gold, would displace more water. Archimedes performed a series of experiments that involved submerging lumps of silver and gold that weighed the same as the crown to see how much water each one displaced. This enabled Archimedes to tell if there was silver in the crown. In his jubilation, Archimedes rushed down the street naked, shouting 'Eureka': 'I have found it'.

Is this the point of physics: to answer seemingly un-answerable questions? We are now able to look at our surroundings across an extraordinary breadth of scales. Where we once thought visible matter to be indivisible, we have gone smaller and smaller, down to the atom, and onward to the most fundamental particles, and ultimately to a view where matter is actually composed of fluctuations in the energy of empty space (see *Are Solids Really Solid?*). The sky was once the limit of our vision; now we know the universe to be so vast that it would take light nearly 28 billion years to cross (see *Am I Unique?*). And, it should not be forgotten, understanding the notion that light has a defined and constant speed is a hard-won triumph of physics, too (see *Can We Travel Through Time?*).

We know much of the history of the universe, the nature of matter and the structure of our planet, but perhaps the greatest lesson we have learned is that, whenever we think we have nature figured out, it surprises us again, revealing just how little we actually know. Isaac Newton probably put it best in his memoirs: 'I do not know what I may appear to the world,' he wrote, 'but to myself I seem to have been only like a boy playing on the sea-shore, and diverting myself in now and then finding a smoother pebble or a prettier shell than ordinary, whilst the great ocean of truth lay all undiscovered before me.'

An alternative to superstition

If there is one aspect of physics' achievements that Newton perhaps appreciated less than most, it was the subject's ability to slice through mysticism and superstition. Newton was a great alchemist and a biblical scholar; he considered his writings on the Old Testament book of Daniel his greatest work. Whenever physics threatened to cast doubts upon spiritual matters, Newton would cringe. 'I have studied these things – you have not,' was his constant retort to astronomers' criticisms of religion. Newton left room for God's work in the mechanism of his 'clockwork heavens' but the march of physics soon displaced the divine hand. When the Emperor Napoleon questioned Pierre-Simon Laplace about his newly published treatise on celestial mechanics, he remarked on the absence of God in the mechanism. 'I have no need of that hypothesis,' Laplace replied. The point of physics, in many ways, is to find what, in the universe, is explicable by a set of laws, and the simpler the laws the better.

Until around 600 BC, civilizations developed technologies but thought little about how to make sense of the world: that was for the prophets and the sages. Then came the Milesians. The city of Miletus, on the west coast of modern

Turkey, was home to a mode of thought that would be recognizable to today's scientists as a thirst for real, first-hand understanding. Rather than having the universe's secrets obscured by mystical religious concerns, the Milesians sought laws to explain the phenomena of nature, and came up with theories for the causes of Earthquakes, lightning and the structure of the universe, among other things.

The Milesians debated these theories openly, considered how they might be tested, and accepted the results of experiments as the arbiter of truth. Anaximenes of Miletus is credited with performing the world's first scientific experiment. His observations of how the temperature of exhaled breath seems to vary depending on whether the lips are pursed or wide open, led him to conclude that compression causes cooling, and expansion causes heating.

The fact that Anaximenes was exactly wrong here is another lesson in the point of physics. It teaches us that we cannot ever be sure of anything that is 'received wisdom'; accepted theories, and even 'facts' about how things in the universe work, are often proved wrong, and supplanted by new ideas. These, too, are open to falsification. Physics is a process of testing everything – especially those things we most want to be true.

It is for this reason that physics is somewhat devoid of 'scientific saints'. It is not so much a discipline of ideas as a discipline of consensus arrived at through the gathering of experimental evidence. Those who fail to accept the results of experiments – and do not provide good reasons why others should join them on the 'wrong' side of the fence – tend to be given short shrift.

More than the sum of the parts

The physicists Albert Einstein and Richard Feynman provide a suitable illustration of the way physics is bigger than any physicist. Though now venerated as a public icon, Einstein did not die a hero to other physicists. On the contrary, his later life is remembered with a tinge of regret at his ultimate quest. Einstein's best-known work was done early in his career. He made a seminal contribution to quantum theory with the experimental discovery of the photon, the quantum of energy (see *What is Light?*).

This destroyed the centuries-old view that light must be a wave. Then his special theory of relativity changed our notion of time. His elucidation of the idea that mass and energy are interchangeable (see *Why Does E = mc²?*) was a revelation about the fundamentals of matter. The general theory of relativity rewrote Newton's gravitational work after nearly four centuries of acceptance (see *Why Does an Apple Fall?*).

After that, though, Einstein's views grew irrelevant to physics. The quantum revolution changed the face of the subject, but Einstein refused to accept quantum theory as a useful way to describe the universe. He spent his later years working, to no avail, on a theory that would unite electromagnetism and relativity and render quantum theory an unnecessary innovation. The number of physicists who would work with him and support him dwindled throughout his life.

Richard Feynman is perhaps the second most famous physicist after Einstein. He was a great popularizer of the subject, a great and innovative thinker, and – most significantly of all – remains a great hero to those working in the field. Feynman never reached Einstein's dizzy heights of achievement, but he did more than most, contributing to the creation of quantum electrodynamics, or QED, a theory that describes the interactions of light and matter (see *What is Light?*). It is widely feted as our most successful theory of physics.

One of Feynman's greatest strengths as a physicist was his ability to listen to the convictions of his peers, bow to the law of evidence, and admit that he was always working from a position of ignorance. He famously said that, 'The first principle is that you must not fool yourself – and you are the easiest person to fool.' His unwillingness to fool himself is summed up in his appraisal of the theory that became Einstein's downfall. 'I think I can safely say that nobody understands quantum mechanics,' he wrote in *The Character of Physical Law*. 'Do not keep saying to yourself, if you can possibly avoid it, "But how can it be like that?" because you will get . . . into a blind alley from which nobody has yet escaped. Nobody knows how it can be like that.'

This is the reason the older Einstein is not revered by

physicists, and Feynman is. While Einstein led himself into a blind alley, Feynman admitted his limited understanding, and followed others as they made forays into new territory. This is another component of the point of physics: progress by building on the achievements of others. As Newton put it, 'If I have seen further it is only by standing on the shoulders of giants.'

Thanks to quantum theory, physics has even taken the extraordinary step of defining some limits for itself. The Heisenberg uncertainty principle (see *Is Everything Ultimately Random?*) sets in stone the fact that there are limits to what physics can tell us about a system.

A humble discipline

When we examine the equations that govern the motion of an electron, say, we can see how they tell us its momentum, or its velocity. There is no means by which they can tell us, precisely, about both the momentum and the velocity, however. The two can be found to only a finite precision.

Werner Heisenberg saw the practical side of this: there are limits to what our experiments can reveal. Bounce a photon of light off the electron, and you can infer its position, but the photon will have imparted some momentum to the electron, too. Thus the act of determining the position of the electron creates an uncertainty in the value of its momentum. Conversely, a measurement of momentum will always create an uncertainty in a particle's position. Whether you look at theory or experiment, there are strict limitations to what we can find out. Physics, in many ways, is a humble discipline. But there's plenty to be humble about, as the physicists behind the atomic bomb will testify.

If you had posed the question 'what is the point of physics?' to Western governments after the Second World War, you would have been greeted by disbelief that you even had to ask. Physics was everything, as the war had shown. Physics had given us fantastic technological innovations: radar, computers, the atomic bomb, and, of course, televisions and microwave ovens. Physics was set to be the driver of economies, and the protector of nations. Pose the same question to physicists, however, and you might have got a rather more subdued response.

Immediately after the first test of the atomic bomb in New Mexico, the Harvard physicist Kenneth Bainbridge turned to Robert Oppenheimer, the project leader. 'Now we're all sons of bitches,' he said. Oppenheimer was dealing with his own mixed emotions: decades later, he admitted they all knew at that moment that the world would never be the same. And yet, Oppenheimer said, put in the same situation, he would do it all again. 'If you are a scientist, you cannot stop such a thing,' he said in his retirement speech in 1945. 'If you are a scientist, you believe that it is good to find out how the world works . . . that it is good to turn over to mankind at large the greatest possible power to control the world.'

The world in our pocket

Is this the point of physics: to gain control over the world? It is true that physics – or at least the industrial application of physics – has created the modern world. If our age can be defined by one thing, it is probably the microelectronics revolution: television, computing, the Internet, and mobile communications, to mention but a few aspects. All of it was built on the back of physics. To be more specific, it was built on the back of silicon technology. During the Second World War, the developers of radar worked to create ever-purer crystals of silicon and germanium for the equipment. Physicists – above all the ones employed by Bell Labs in the USA – continued that development after the war, learning how to turn them into 'semiconductors' and incorporate them into technologies that had previously required in-efficient and bulky valve amplifiers. By 1952, the first silicon-based electronics products had hit the market: low-power and highly portable devices, such as hearing aids and pocket radios. A year later, the first transistor-driven computer appeared. Shortly after that, people started to refer to the concentration of electronics companies in a small area of northern California as 'Silicon Valley'.

It is not hard to see the impact of physics on our lives. Lasers provide a specific example. Lasers also came from Bell Labs, and stemmed from wartime research into radar technology. Since their invention in 1957, they have become ubiquitous in everyday life. CD and DVD players, fibre-optic communications systems such as the telephone network, supermarket checkout scanners, eye surgery and laser printers are just a few of the applications.

So, is the development of technology the point of physics? Not at all. The technological revolutions of the 20th century came about as a result, ultimately, of the discovery – or invention, if you prefer – of quantum theory. That was the result of trying to unravel things no one understood, such as why the spectrum of radiation emitted by an oven at 100 celsius was the same as the spectrum of radiation emitted by anything else at 100 celsius, rather than specifically trying to invent new devices.

In essence, our modern electronic technologies come from quantum theory, which came from thermodynamics, the study of heat. That arose from the study of gases – and so on. Physics is a self-sustaining chain reaction: every discovery provokes another set of questions, which provoke new discoveries. As George Bernard Shaw once said, 'science never solves a problem without creating ten more.'

A never-ending story

There is no end of questions in sight. Physicists used to be fond of saying their work was done. In 1894, the American physicist Albert Michelson announced that, 'The most import-ant fundamental laws and facts of physical science have all been discovered, and these are now so firmly established that the possibility of their ever being supplemented by new discoveries is exceedingly remote.' Within a decade, we had the twin revolutions of relativity and quantum theory.

In 1888, the astronomer Simon Newcomb had announced the end of astronomy: there was little left in the heavens to discover, he suggested. Newcomb was wrong too. Our view of the cosmos has probably changed more radically since Newcomb's time than it did in the thousands of years of scientific discovery that took place before he was born. Although the major breakthroughs of the last century showed us where we came from, outlining the entire history of the universe, the hubris is gone from our world view; with the discovery that most of the universe is in a form unknown to science, physicists now appreciate that they have got to grips with only a tiny percentage of the universe.

There is, it has to be said, one end in sight: the theory of everything. If physics began with the Milesian quest for the laws governing natural phenomena, it will (theoretically) end with the discovery of just one law: the ultimate description of the universe. This 'theory of everything' will reduce all the particles, the forces that govern their interactions and the space and time in which their existence plays out, to a single unified description (see *Is String Theory Really About Strings?*).

At the moment, we are far from achieving that goal, but here, perhaps, we have found the true point and the essence of physics: to discover the span of our ignorance, and to do what we can to reduce it. Sometimes, as with the atomic bomb, there is a price to be paid for this journey of discovery. Sometimes, as with the development of quantum mechanics, we reap great practical rewards from it. But most of the time, physicists will tell you, physics is simply about the thrill of discovery – and then discovering that our discoveries have made the world more interesting, not less. As the poet John Dryden said, 'Joy in looking and comprehending is nature's most beautiful gift.'

What is time?
Progress, disorder and Einstein's elastic clocks

Deep in your brain there lies a lump of tissue called the striatum. This assortment of neurons is, to the best of our knowledge, the only dwelling place of time. It accumulates the first record of the moments of your life, and provokes your sense that your childhood was a tumbling assortment of significant and fascinating moments, while adult life hurtles by too fast to be properly appreciated.

You shouldn't set too much store by these sentiments, though. The striatum's gift is actually to create an impression – perhaps even an illusion – of time passing. The problem is, its measure of time depends on what is going on in your conscious mind. Every time you perform a conscious task such as putting the kettle on, the various electrical circuits in your brain spike in unison. The striatum records this simultaneous signalling and starts to note the subsequent patterns of electrical signalling from areas such as the frontal cortex. Your notion of how much time has passed before the kettle boils is nothing more than a measure of the accumulated electrical signals.

That's not so bad at home, where you can calibrate it with a glance at the kitchen clock. But as soon as you are denied access to clocks, things go awry. When, in the early 1960s, the French geologist Michel Siffre took off his watch and lowered himself into a dark cave for 60 days, his perception of passing time unravelled. By the end of the experiment, what Siffre thought was an hour was often four or five. Drugs such as valium, caffeine or LSD will send your sense of time similarly awry. As will your memory.

We often think busy times make life flash by, but experiments show that's only true *while* you're busy. Afterwards, when you reflect on your existence, your busy periods will seem much longer. That's why your childhood now seems to have been a series of long, golden summers – life was exciting when you still had so much to experience, and your brain thinks that those heightened signalling levels must correspond to huge stretches of time. Your grip on the passage of time, then, is as precarious as you may always have suspected. But it turns out that our problems with the perception of time are as nothing compared with our problems with the notion of time itself.

Universal time

You might think we ought to have a handle on time by now. After all, time is a universally understood concept – every human culture knows about it, talks about it, feels it. And we have been thinking about what it means for millennia. In 350 BC, Aristotle, for instance, wrote a work called *Physics*, which included one of the first attempts to grapple with the notion of time.

Aristotle's work on time begins with a question. 'First,' it says, 'does it belong to the class of things that exist or to that of things that do not exist?' Here in the second millennium AD, that is still an open question. If our minds are fooled by the passage of time, that may be because time itself is an illusion. From the Greeks to modern-day physics, the main conclusion about time has remained constant: it is, at the very least, about change. Through time, one thing changes into another.

But while Aristotle's Greek peers were obsessed with the circle as the most fundamental concept in the universe, meaning that time must flow in cycles, modern physics is focused on linear processes: beginning to end, Big Bang to cosmic shutdown. With time, that translates into an overwhelming sense of time's arrow: in our modern view of the universe, time moves irreversibly forward. Eggs break, and cannot be unbroken. Clocks wind down, and do not spontaneously wind up.

This process of change, in which systems move irreversibly into disorder, is known as the thermodynamic arrow of time. It arises from one of the most fundamental laws of physics: the second law of thermodynamics. This states that, as a whole, the universe is caught in a process of unravelling order. Entropy, a measure of the disorder in a system, is always increasing.

Order and disorder

The arrow of time might arise from a variety of sources. The 'cosmological arrow of time', for example, cites the creation of the universe as a move away from a special, low entropy state where everything was neatly ordered. It is rather like handing a fully solved Rubik's Cube to a curious child; as time progresses, the universe moves to an ever-more disordered state, just as the neat order on the faces of the Rubik's Cube will give way to a messy jumble of colours. While some things, such as galaxies, appear ordered, with structures that are often intricately beautiful, the order of the universe as a whole is decreasing. The end will come when there is no more disorder to be created; or, as Lord Kelvin put it, when the universe has reached 'a state of universal rest and death'.

Our familiar arrow of time could equally result from quantum theory. In one (probably the most popular) school of thought, quantum systems undergo an irreversible 'collapse' when they are measured. This originates from the remarkable ability of a quantum object such as an atom to exist in two entirely different states at once. It might, for example, be spinning clockwise and anticlockwise at the same time. When the measurement is made, however, that double state is forced to become one or the other: the measured atom will be found to be spinning clockwise or anticlockwise, and will not spontaneously revert to the state of doing both.

There is a problem with these descriptions of time's arrow, however. They get us nowhere because they require the concept of change. And change, as Aristotle noted, is a marker of time passing. Through considerations of the arrow of time we are really no further forward in defining time. All we have is a putative explanation for the direction it appears to take. And even that has been undermined. Time's arrow might be part of our individual experience, but we have no reason to believe that makes it real. Worse still, we have good reason to believe it isn't.

A stretch in time

We have Albert Einstein to thank for this alarming insight: it lies at the heart of his special theory of relativity. Einstein was relatively unknown when he published his ideas in 1905. Special relativity was a revolutionary work, dismissing in a single stroke the popular and long-lived concept of the ether, a kind of ghostly fluid that fills all of space and provides a background through which electromagnetic fields such as light could move.

It is worth mentioning at this point that while, as the late Carl Sagan once said, extraordinary theories require extraordinary evidence, special relativity is one of the few such theories where extraordinary evidence has been found to back it up. What you are about to read may seem absurd, but there is every reason to take it seriously.

The central point of special relativity is that the laws of physics work the same for everyone, regardless of how they are moving through the universe. The most important consequence of this is that the speed of light is a constant, universally known as c. If you were to measure the speed of the light emitted from the headlights of a vehicle travelling towards you at 100 kilometres per hour, the speed of the light would be c, not c plus 100 kilometres per hour (62 mph). The speed of light does not change depending on the relative motion of the emitter and observer. The extraordinary upshot of the constancy of c is that, when conditions require it, everything else does change – and that includes time. The passage of time is as flexible an affair in the real, physical world as it is inside your mind.

Let's imagine a scene where you are standing 100 metres from an intersection controlled by traffic lights. You are equipped with a stunningly accurate stopwatch, a metre rule and lightning reflexes. The light changes to red, and you are able to measure the time it takes for the first pulse of red light

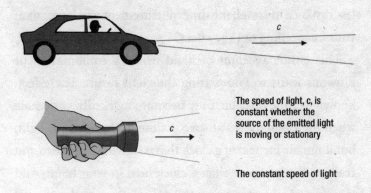

The speed of light, c, is
constant whether the
source of the emitted light
is moving or stationary

The constant speed of light

to travel the length of your metre rule. At that moment, a car
passes you, travelling towards the intersection at 100 kilometres per hour. The passenger in the front seat has the same
skills and equipment as you, and makes the same measurement: the time taken for the light to travel the length of
the ruler.

You have both measured the speed of light, and Einstein
insists that you must both get the same result. But as the car
moved past you towards the traffic light, the metre rule
within it also moved past you. By the time the light reached
the end of the ruler in the car, the far end of the ruler was
closer to the traffic lights, and so the light had to travel less
distance compared with yours. The passenger in the car
should measure light as faster, completing a metre in less
time. How then, can you both get the same result? The
answer has to do with the passage of time in different
situations. Compared with your clock, the clock in the
moving car runs slow. So, although the light apparently had

less distance to travel, the time measurement was larger than yours, cancelling out the effect.

This is not a sleight of hand where a combination of illusions leads to you getting the right result. The effect, known as time dilation, only becomes markedly noticeable when the clock moves at speeds close to the speed of light, but it remains true that a clock that is moving relative to you really will run slower than a clock held in your hand. And the word 'clock' refers to anything that can mark the passage of time. Dissect that statement, and you'll find that all kinds of disturbing implications emerge.

Ageing relatives

Let's start with something that is just about conceivable. Take a lump of polonium, a radioactive material discovered by Pierre and Marie Curie around 100 years ago. One form of polonium, polonium-209, has a half-life of about 100 years; that is, after a century, half of its atoms will have emitted a burst of radiation and transmuted into more stable atoms.

If the Curies had taken two identical lumps of this material when they discovered it, and left one in their Paris laboratory while shooting the other one on a round trip into space at 0.99 of the speed of light, returning to Earth today, we would notice something remarkable about the amount of radiation they were giving off. The lump that stayed in Paris would lose half of its radioactive polonium atoms during that century. The thing is, its twin, the lump that had rocketed into space and back while 100 years passed on Earth, would only have lost 10 per cent of its radioactive polonium atoms.

That is because the motion relative to Earth at 0.99 the speed of light (setting aside practical issues such as acceleration, deceleration and turning round) slows time for this lump. Its 'clock', as measured by the rate at which its atoms experience radioactive decay, is running at only 14 per cent of the speed of its twin that never left the planet. That is why so many of its radioactive atoms remain intact. This, perhaps, is hard enough to swallow. But now for something truly inconceivable.

Let's allow Pierre and Marie Curie to guard the two lumps of polonium. Pierre will accompany one lump on that same return trip into space, while Marie remains in Paris with her lump. The scientists' bodies have internal clocks, too: as with the polonium, their atoms change with the passage of time, creating a heartbeat, for instance, and cells that shut down after performing a certain number of divisions – a phenomenon that biologists believe to be the root of ageing and death.

Turning a blind eye to the likely catastrophic effects of the radiation, the atoms – and thus the cells and the heartbeat – in Pierre's body will run slow compared to Marie's, just as the polonium's radioactive decay runs slower than on Earth. When Pierre returns, 100 Earth years later, Marie is long dead, but Pierre's body has aged only 14 years. One immediately obvious conclusion from this is that, given the right resources, time travel into the future is entirely possible. But it is a short step from this point to the astonishing revelation that Einstein's special theory of relativity does away with the notion of some common future anyway. And neither is there a common present or past.

In search of lost time

You might claim, as you stand looking at the traffic lights, that you saw two events happen simultaneously. But as we have seen, the passenger in the car has a clock that runs at a different speed. The information they gain about the timing of those two events could well be different. Worse, you might see two events, A and B, happening at distinct times, with B following A. Depending on how your relative friend is moving, however, they could see A follow B. That is potentially catastrophic: if you think A caused B, how is that explicable to someone who saw B happen first?

Past, present, future, simultaneity, cause and effect – nothing is universal. When it comes to time and the processes it governs, you and your striatum really are on your own. There is a simple answer to all this confusion, however, and it is an answer that is appealing to many physicists and philosophers. We could do away with the very notion that time exists.

It is an argument that harks back to the 17th century. Newton, whose Christian faith required that space and time reflect the character of God, considered time to be a real entity, an absolute that moves on independently of everything in the universe. But his great rival Gottfried Leibniz believed time to be a human construct. All we can do, Leibniz said, is describe how the positions of things in space relate to each other, and how that relation evolves. It is useful that a clock's pendulum swings back and forth and the clock's hands circulate around the dial in response, for example, but that doesn't mean the clock is measuring something that actually exists. Time, in this view, comes out of our desire to make sense of the world, but it is no more than a useful means of orientation. It is a shorthand, like the spatial concept of 'up'. 'Up' means a certain direction when I am stood in London, but the same direction is actually 'down' in Sydney.

This link is slightly more than a convenient illustration. When Einstein published his general theory of relativity (the 'special' in 'special theory' refers to a special, i.e. particular case, not a special significance), he postulated a bond between time and space. Time, he said, is just one of four dimensions to the universe. The other three are the familiar ones in which you move your physical body: up and down, across, forwards and backwards. The only difference is that, while we conscious creatures can choose how we move through the spatial dimensions, we have no control over our movement through time.

Stretching space and time

Einstein's four dimensions of space and time – together known as space–time – can be thought of rather like a piece of fabric that can be distorted, bent, folded, twisted and even torn by anything within them that has mass or energy. From this foundation, general relativity has equipped us with equations that describe the features of the cosmos with unprecedented accuracy, allowing us to find out how the universe works, send spacecraft to distant destinations and create the array of global positioning satellites that tell us where on Earth we are. But perhaps most intriguingly of all, the pliable nature of Einstein's four-dimensional fabric hints at the origin of time.

Your mass distorts space–time very little. The mass of the sun distorts it much more – according to general relativity, this distortion is the root of the gravitational attraction that keeps our planet in orbit. Even more powerful is the distortion that is brought about by a collapsed giant star: a black hole. And it is here that we glimpse the true power of Einstein's work.

The enormously strong gravitational field of a black hole means that there is a spherical region close to its centre where the velocity required to move away from the black hole is greater than the speed of light – an impossible velocity to achieve. Nothing, including light, can get out of this region, and so we cannot gain any information about anything that goes on beyond its boundary. Hence its name: the event horizon.

At the event horizon, time dilation is infinite. Somebody watching from a safe distance as you fall towards the event horizon would see your movements slow down then freeze as time runs infinitely slowly for you compared to the observer. Only in the observer's infinite future would you reach the event horizon, so you never actually disappear from view. Your experience, on the other hand, would be hugely dramatic. Your body is extremely unlikely to survive the enormous gravitational forces, but if you did survive you would eventually encounter what, according to relativity, is a breakdown in the very fabric of space–time. This 'singularity' at the centre of a black hole occurs as the distortion becomes infinite. Here, we reach the limit of the known laws of physics – beyond this point, they no longer apply.

The moment when moments began

Though it is commonly associated with destruction, the singularity is also thought to be the key to creation. In the early 1970s, Roger Penrose and Stephen Hawking adapted the mathematical notion of the black hole singularity to explain the origin of the universe. In a black hole everything disappears into the singularity. Reverse the mathematics of the process, though, and the singularity could give birth to the very fabric of space–time. For more than three decades this has been seen as our best description of the Big Bang, the origin of time itself.

If general relativity sheds some light on where time comes from, it still does not tell us a great deal about what time is. What's more, impressive as Einstein's formulations of the character of space and time are, we know that special and general relativity are not the final answer.

If the singularity shows us anything, it is that, while general relativity works remarkably well in many scenarios, it offers no satisfactory explanation for the most extreme phenomena of our universe. A more complete description of the cosmos and how all its contents (including the centres of black holes) behave – a theory often referred to as 'quantum gravity' – still eludes us. And, as it turns out, the nature of time is right at the heart of the problem.

Quantum gravity has to work relativity's notions of time into quantum theory, our best description of how the microworld of molecules, atoms and subatomic particles behaves. But quantum theory takes little note of time. In the standard formulation of the theory, you can't ask questions about how long a process takes, for example. Then there's the problem that quantum theory tells us that most of the sub-atomic particles exist independently of the direction of time. Just as they can spin clockwise and anticlockwise at the same time, their quantum states can evolve forward and backward in time. Researchers are even learning to do quantum experiments where information seems to come from the particles' futures. What's more, special relativity tells us that massless particles, such as photons and the gluons that bind nuclei together, travel at the speed of light and do not even experience the passage of time.

The great physicist John Wheeler once said, 'Time is nature's way to keep everything from happening at once.' He would have said it with a twinkle in his eye, knowing full well that the apparent simplicity of time belies its true nature. Saint Augustine was more honest when he said, 'What then is time? If no one asks me, I know what it is. If I wish to explain it to him who asks, I do not know.'

Despite all our scientific achievements since Augustine, time remains an enigma, possibly the biggest question facing physicists today. But if time is an illusion, it is at least a useful one. Our interpretation of its consequences – our memories of the past, our existence in the present and our hopes for the future – lie at the core of the human experience. Or that's what your striatum wants you to believe.

What happened to Schrödinger's cat?

Quantum physics and the nature of reality

It was 1925, the heyday of Buster Keaton and Charlie Chaplin. The world was getting excited about *The Gold Rush*, hailed as Chaplin's finest film to date, coming out next month. And poor Wolfgang Pauli, a physics student based in Hamburg, Germany, was depressed. 'Physics at the moment is again very muddled; in any case, for me it is too complicated,' he wrote to a colleague. 'I wish I were a film comedian or something of that sort and had never heard about physics.'

Pauli was right: physics was muddled. No one understood what the newly formed quantum theory was all about. Experiments dictated that energy must be split into indivisible packets or quanta, but no one could say why. Then, just a few months later, the Austrian physicist Erwin Schrödinger cleared the confusion. It happened during a trip into the Swiss mountains with a woman who was not his wife, and ended with his questioning the fate of an imaginary cat. The creature quickly became the most famous animal in science.

The story of Schrödinger's cat has the weirdness of quantum theory running right through it, and its appropriately enigmatic nature remains intact to this day.

The source of Schrödinger's breakthrough lay in the work of a French physicist called Louis de Broglie. In 1923, de Broglie put together relativity, generally the physics of the very largest scales of distance and speed, and the nascent quantum theory, the physics of the very small. The outcome was a simple equation. Every moving particle, de Broglie said, could equally well be described as a wave. Every wave could be described as a moving particle. Einstein, when presented with the work, pronounced it 'quite interesting'. Two years later, however, Schrödinger showed it was much, much more than that.

Erwin Schrödinger worked out the mathematical implications of de Broglie's formula during a Christmas holiday in 1925. Leaving his wife in Zürich, Schrödinger took his mistress off to a chalet in the Swiss mountains. It was not unusual behaviour for him – he and his wife seemed to come to several 'arrangements' through their marriage. Whatever went on, the trip was obviously inspiring. Schrödinger came back from the mountains with what is now known as the Schrödinger wave equation. This describes how a quantum particle behaves when it is considered as a wave.

The Schrödinger equation provides a way of understanding where quantum states come from. Take the Bohr model of the atom, for example, where an electron circling the nucleus can only have particular energy states. Schrödinger's equation gives a way of working out what those 'quantized' energies are: the electron is stable only when its wave completes a whole number of oscillations during its orbit.

This was a revelation to physicists, who had no proper justification of the quantized energies. But the equation also gives a way of working out how the energy, say, of an electron will evolve over time in a particular situation. It can equally well give us the particle's position, or its momentum, or how the quantum states of two interacting particles will end up. It was hailed as a masterstroke. There was only one problem.

No one could agree on what the wave equation actually meant. Did it mean the particles were really waves? Schrödinger believed – or rather hoped – so. Einstein stood with him. But others disagreed. The University of Göttingen physicist Max Born, for instance, showed that solutions of the wave equation might give nothing more than probabilities. The probability of finding a particle in a particular space, say, or the probability that a particle will have a certain momentum.

In this view the equation was a guide to what we might find out about the quantum system under inspection, but had nothing to say about what the nature of the system actually was. In other words, it did not give us a description of the quantum object, only a description of what we could know about it. Philosophically, this was a nightmare. Einstein hated it, as did Schrödinger.

Positive thinking

Niels Bohr, on the other hand, loved it. Bohr was based in Copenhagen, where he ran an institute sponsored by the Carlsberg brewery. He was a 'positivist': his philosophy said that it was meaningless to talk about something's objective properties because you could only ever access knowledge about it through subjective measurements. Those measurements will always impose restrictions on what we can know.

The ultimate reality behind Schrödinger's wave equation was neither wave nor particle, Bohr felt, and so could not be described in any terms we can deal with. His answer was to assume that nothing exists until it is measured. But once a measurement is made, the type of measurement will determine what we see. If you use an instrument that detects something's position in space, for instance, you'll see something that has a definite position in space – the entity that we call a particle.

Einstein would have none of this 'Copenhagen interpretation' of quantum theory. His great work, relativity, had been built specifically to create a theory that was independent of the observer. The central theme of relativity was that the laws of physics should be the same, whoever is working them out. The notion that the physical nature of the universe was dependent on how we looked at it offended his sensibilities deeply.

Einstein's problem lay in the fact that describing quantum objects using a wave equation meant that, like waves, they could interfere with one another. When two waves interact, they produce a 'superposition', which is the sum of the waves at any point. Where two crests coincide, the superposition is larger than both. When two troughs coincide, the wave trough deepens. If a crest and a trough coincide, the result is flat.

How does this apply to quantum particles? Schrödinger's wave equation says that, in the right circumstances, they exist in a superposition of different states. Thus an electron circulating in a ring of metal can be circulating clockwise and anticlockwise at the same time. A photon of light can be polarized – that is, have its electric field oriented – in any number of directions at the same time. A radioactive atom, which decays via a quantum process, can be in a superposition state of 'decayed' and 'not decayed'. Though it seems nonsensical, this is what the theory states.

Which is why Einstein and Schrödinger said there must be something missing from the theory. And, to hammer his point home, Schrödinger came up with the cat. 'One can even set up quite ridiculous cases,' Schrödinger wrote in a 1935 journal article. 'A cat is penned up in a steel chamber . . .' Schrödinger went on to describe this 'ridiculous' case in some detail, unwittingly creating the touchstone for future interpretations of quantum theory.

Cat in a box

In the closed steel chamber with Schrödinger's cat is a tiny piece of radioactive material and a Geiger counter. At any moment, there is some probability that the radioactive material will emit a particle, thus triggering an electrical current in the Geiger counter. But Schrödinger had the Geiger counter rigged up to release a hammer that, on sensing a radioactive emission, would smash a flask of hydrocyanic acid, releasing vapours that would kill the cat.

According to Schrödinger the quantum description of the entire system, including all the atoms that make up the cat, 'would express this by having in it the living and dead cat (pardon the expression) mixed or smeared out in equal parts.' The logic is sound. The indeterminate nature of the radioactive atom, in a superposition of 'decayed' and 'not

Schrödinger's cat thought experiment

decayed' can also put the cat in a superposition of dead and alive.

The kicker comes when the issue of measurement is brought to bear. Bohr had said that there is no definite reality until a measurement is made, because the choice of measuring instrument determines which facet of the system – wave or particle, for instance – the observer will see. So, in Bohr's view, the act of opening the box and observing the state of the cat would force it to be alive or dead.

This was what Schrödinger found so ridiculous: how can the act of observation change such a fundamental property of a cat? It must be one thing or the other; Bohr was being fooled in the same way that a blurred photograph can give an impression of fog, he said. 'There is a difference between a shaky or out-of-focus photograph and a snapshot of clouds and fog banks.'

By this time, though, the interpretation of quantum theory was already a matter of public debate: Einstein and Bohr had a famous exchange in 1927, at the fifth Solvay Conference in Brussels. Einstein challenged Bohr with a series of thought experiments. Imagine such and such a situation, he would say: how can the observation, or the interaction with the apparatus cause a superposition to resolve into one state or the other?

Waves and bullets

The eventual outcome of this argument was a new version of
an old experiment: the famous 'double slit' experiment. In
1801, Thomas Young overturned Newton's particle view of
light by shining light at a screen scored with two slits. Young
observed an 'interference' pattern, which can only be
explained through superposition of waves. The quantum
version asks what happens when you reduce the light intensity so far that quantum theory kicks in. When there is
only one bullet, or 'photon' of light in the experiment at any
one time, there can be no interference, surely?

In Bohr's view there could – as long as no one was looking
to see which slit the photon travelled through. To Bohr, the
light is neither a wave nor a particle – those are names that
we give something whose properties we have measured.
According to Schrödinger's wave equation, the photons of
light go through both slits. Despite being a single particle,
each photon is 'smeared out' as a wave, effectively having
two independent existences as it passes through the slits. As
long as no one measures the path the light takes, it takes all
available paths.

You might think that this is all wordplay – abstract
thought experiments whose weirdness will disappear once
the experiments are carried out in the real world. You would,
to Bohr's delight, be wrong. We didn't find that out for sure
until relatively recently. The first double slit experiment with

only one particle in the apparatus at any one time was only carried out in the 1970s. But it worked: despite being faced with two slits, a succession of electrons gradually built up an interference pattern on the screen beyond the slits.

And spookily, when an instrument was placed in the experiment to measure which slit the electron went through, the interference pattern disappeared. In other words, measurement made it manifest as a particle, not a wave. That might seem far removed from Schrödinger's cat – a cat, after all, is a very different beast from an electron. But subsequent experiments have pushed the quantum particle to ever-larger sizes.

We have carried out the quantum double slit experiment with photons, electrons, atoms, and even 60-atom fullerene molecules. The weird interference effect has never disappeared – unless we tried to look at which slit the particle went through. Plans are afoot to do it with much larger objects: a virus, and maybe something a million times bigger than the fullerene molecule. Apart from the difficulties of building the experiment, there is no fundamental reason to stop there: there is no cause to suggest why a real cat shouldn't behave in the same way as an electron, given the right circumstances and a cat-flap-sized double slit.

Except, of course, that it is easy to see a real cat, and thus determine which cat-flap it went through. In Schrödinger's thought experiment, the box has to remain closed so that no one can see the cat, there is no measurement performed, and the superposition remains intact. This leads us to a difficult question, one that Bohr always evaded. What constitutes a measurement? With Schrödinger's cat, is it when the box opens? When light photons bounce off the cat relaying to us the information that allows us to tell whether the cat is dead or alive? Or is it when those photons enter our eyes? Or when our conscious minds register the state of the cat? Bohr's answer to this conundrum was, essentially, that physicists just know when they have made a measurement. Modern versions of the Schrödinger's cat experiment, however, are shedding much more light on the process – and explaining why a cat can't really be dead and alive at once.

Don't look now

The boundary between the 'classical' world that we inhabit and the quantum world of the atoms comes down to the de Broglie waves that brought this whole story into existence. The de Broglie wavelength of a body, which depends on its momentum, gives a measure of the scale at which it will manifest as a quantum wave.

In the double slit experiment, the fullerene molecule has a de Broglie wavelength of around 10^{-12} metres, or a thousand billionth of a metre. The gap between the slits is around half a million times bigger than that; bigger, but not too different in scale. This means the system is suited to exposing wave behaviour. This is still in line with Bohr's claim that the choice of measurement apparatus decides which characteristics will manifest, but it does throw out two explanations for why a cat or a person can't – unlike the fullerene molecule – seem to be in two places at once.

The first reason is practical. Walking along a wall at a couple of miles per hour, for example, Schrödinger's cat would have a wavelength of around 10^{-28} metres. Its quantum, wave-like behaviour would only be exposed by a measuring device of a similar scale. Since we have never created such a device, we cannot perceive quantum behaviour. Everyday life is, according to Bohr's scheme, an experimental situation that will always manifest the particle-like nature of everything around us.

The second reason that we are 'classical' is that we are emitting radiation. Anything that has a temperature above absolute zero, −273 celsius, emits photons, packets of energy that carry away heat. Experiments have shown that this radiation can be used to find the location of the object, effectively revealing which slit it passed through. In other words, at a temperature above absolute zero, you can't close the box on Schrödinger's cat, invalidating the premise of the thought experiment whenever you translate it into the real world.

These experiments were carried out by firing fullerene molecules at a double slit. The hotter the fullerene molecule was as it approached the slits, the more blurred the interference pattern. The hot molecule emits photons, and the energies of the emitted photons are determined by the temperature. Higher temperature essentially gives higher energy, which translates, in de Broglie's terms, to a shorter wavelength. And the shorter the wavelength of the emitted radiation, the easier it is to infer the emitting molecule's position. In other words, a hot body seems to give away more information about which slit it might go through.

The same thing happens if the fullerene molecules collide with air molecules on the way to the slits. Normally the experiments are done in high vacuum, but if the vacuum is not so good, and the position of the fullerene can be inferred by watching what it does to air molecules, the interference pattern fades away. Again, as it becomes possible to infer which slit the molecule goes through, its ability to go through both at the same time begins to disappear. In a partial vacuum, the fullerene behaves as if someone had left the box half-open on Schrödinger's cat, forcing it to be alive or dead, but not both.

So, information does not have to enter a conscious mind to constitute a measurement: it just has to leak away from the system under scrutiny. It appears that a flow of information about the health of Schrödinger's cat is enough to force it into one of the two states available. Where humans and cats are concerned, that information leaks away because our bodies interact with our environment in myriad ways, radiating heat and knocking air molecules about. Information about the position of our bodies is available, which means we can't be in two places at once. This spilling of information is known to scientists as 'decoherence'. Decoherence is not a trivial issue: it might just show us the very nature of the universe.

Information and reality

The physicists looking into the enigma of Schrödinger's cat are now wondering whether it points to the notion that information is the most fundamental element of reality. Quantum theory, in the form of Schrödinger's unfortunate cat, suggests that the universe can be described as one giant information processing machine. And this leads to potential applications too. The role of information in quantum theory has led us to one of our most ambitious technological projects: the quest to build a super-powerful processor called a quantum computer.

The idea behind a quantum computer is to use the Schrödinger's cat phenomenon to perform computations on a massive scale. Familiar computers use the charge state of a capacitor to represent a number in binary: 0 or 1. Quantum computers, on the other hand, use the state of an atom. If it is in its normal state it is 0. If it is given a little extra energy, it is 1. But, being a quantum object, the atom can be in a superposition of 0 and 1 at the same time.

Using another quantum phenomenon called 'entanglement' to string together lots of atoms in superposition allows quantum computing researchers to create a string of undetermined numbers that, when put through a series of steps, perform computations on all possible numbers at once. Quantum computing is a way of doing 'parallel' computations on an unprecedented scale. In theory, an entangled string of just 250 atoms, each in a Schrödinger's cat superposition state, can encode more numbers than there are atoms in the universe. The potential is huge. No wonder governments are seeking to protect their national security ciphers from the developers of the first quantum computer.

There is just one problem. The nature of entanglement and superposition make the atoms especially vulnerable to losing information, and when they do, the computation falls apart. If researchers could get more of a handle on decoherence, and why we never see alive-and-dead cats, they might be able to usher in a revolution in computing.

However powerful it turns out to be, the quantum computer is unlikely to be able to help us comprehend how a cat really can be alive and dead at the same time. The idea that this is part of the nature of physical reality remains truly outrageous to the human mind. Wolfgang Pauli, who didn't give up on physics, and became one of the most brilliant physicists in the history of science, was right. It's too complicated to grasp. As Niels Bohr once said, 'Anyone who is not shocked by quantum theory has not understood it.'

Why does an apple fall?
Gravity, mass and the enigma of relativity

Because of gravity, of course. Everybody knows that. However, what is the fundamental nature of gravity? That is a much harder to question to answer, despite the fact that gravity is the first of nature's fundamental forces to penetrate your consciousness.

Here's an experiment you can try at home. You'll need a six-month-old baby (you could borrow one). Tie a piece of fishing line to one of the baby's toys – a rattle, say. Now suspend it from the ceiling at a height where it will rest lightly on a chair with the line taut and invisible. Get the baby to watch as you whip the chair away. Keep your eyes on the baby's: when, for no obvious reason, the rattle doesn't fall to the ground, the baby will stare at it for much longer than is normal.

This, according to psychologists, is how babies express astonishment. It seems that we know from a surprisingly young age that things are meant to fall downwards when unsupported, and we are mystified if they don't. No wonder the levitation tricks of the Victorian illusionists entranced an entire generation. When things cheat gravity, our very core takes delighted offence.

Gravity, you see, is a tyrant. It cannot be cheated. We cannot, as we can with an electric or magnetic field, block it out. Neither can we counter it with an opposing force – there seems to be nothing in physics that equips us with anti-gravity. The rule of gravity is so central to human experience that we have become, essentially, oblivious to gravity's presence. It is only in its absence – or, rather, its apparent absence – that we remember it is always there.

Perhaps that is why the earliest ventures into science largely ignored gravity. As we understand it now, one type of action governs the fall of a tripping human, the arc of an arrow's flight and the motion of the planets, but Aristotle's textbook *Physics* makes no mention of any universal force orchestrating the universe. He did suggest that objects did not fall off the Earth because of the Earth's 'heaviness', but his reasoning was askew. He suggested that the strength of the Earth's pull depended on how big an object was and what it was made of.

In Aristotle's view, heavy objects fall more quickly than light objects. That is because of the Greek obsession with the elements: Earth, Air, Fire and Water; most of the heavy objects Aristotle knew about were made from materials found in the Earth, and the strong pull, he said, was because they were compelled to return there. Our understanding didn't really move on from this flawed idea for almost 2,000 years. Eventually, though, the Italian scientist Galileo Galilei established that Aristotle was wrong: heavy objects are not more strongly attracted by the Earth. As long as air resistance is not a factor, a heavy and a light object will fall at the same rate.

As easy as falling

Sadly, the romantic stories about Galileo's proof of this – by dropping cannonballs from the leaning tower of Pisa – are not true (the myth was started by Galileo's student Vincenzo Viviani), but it has nevertheless been proved in an even more spectacular fashion. In 1971, *Apollo 15* astronaut David Scott paid tribute to the discovery's profound consequences by dropping a hammer and a falcon feather onto the moon's surface. 'One of the reasons we got here today, was because of a gentleman called Galileo,' Scott said as he let them fall. The hammer and the feather landed, of course, at the same time.

Scott's appraisal was almost correct: astonishingly, it really didn't take much more than Galileo's 17th-century insights to get us to the moon. The gaps were filled by a man born just one year after Galileo died: Isaac Newton. Unimpressive as he was at birth – his mother said he could be 'put in a quart mug' – Newton took just a couple of dozen years to gather all the information it would take to plot a course for the *Apollo* astronauts four centuries later. And here, of course, is where the apple comes in.

Unlike the stories of Galileo's experiments on the leaning tower of Pisa, accounts of Newton's gravitational epiphany at the sight of a falling apple are almost certainly true. It was late summer, 1666, and Newton was sat in his garden at Woolsthorpe Manor in Lincolnshire. The apple tree is still there, and still bearing fruit every autumn.

An apple falls because it has a property called mass, and so does the Earth. Newton's great leap forward was to spell out how everything with mass attracts everything else with mass. His universal law of gravitation, constructed at the tender age of 23, said that the attractive force is dependent on those two masses, the distance between them, and a constant known as G.

Actually, physicists are often over-familiar with the gravitational constant and call it 'Big G' to distinguish it from (little) g, the acceleration due to the Earth's gravitational pull. However, despite the familiarity, G is actually the least well defined of all the fundamental constants.

The size of G, like that of all the other fundamental constants, is known not through some theoretical argument, but through measurement. The English physicist Henry Cavendish was the first to measure it, in 1798, by analysing the gravitational attraction between two known masses that were a known distance apart. His answer for G was 6.754×10^{-11} metres cubed per kilogram per second squared. Today, G is officially 6.67428×10^{-11} m3/kg/s^2. The uncertainty on this measurement is about one part in 10,000. Compare that to the precision with which we know the other fundamental numbers, such as the Planck constant used in quantum theory: that is known to 2.5 parts in 100 million.

There are two reasons G is so difficult to measure accurately. The first is that it is impossible to screen out gravitational fields using any known physics. That means any measurements have to take into account the influence of any and all objects in the vicinity. This makes the measurements unreasonably sensitive to external influence; there are stories of researchers having to recalibrate their apparatus after someone two laboratories away has moved a large pile of books into their office. For this reason, gravity measurements have to be done in isolated laboratories using extraordinarily sensitive instruments.

The second difficulty with measuring the gravitational constant is the fact that gravity is the weakest of the fundamental forces. When that apple falls to the ground, it does so with relatively little acceleration, despite the fact that the mass of the entire planet is tugging it downward.

If you're not convinced that gravity is weak – maybe you've done a parachute jump or been on a rollercoaster and experienced a terrifying acceleration – think about the magnets sitting happily on the door of your refrigerator. The mass of the entire planet is working to pull them towards the ground too – and yet a button-sized dot of magnetized iron can resist the planet's pull. Magnetism results from the electromagnetic interaction between charged particles inside a magnet. And that force is around 10^{42} – that is, around 1 million trillion trillion trillion – times larger than the gravitational force between them. So gravity is weak: G is astonishingly small. But why? Though the weakness of gravity is one of the central mysteries of physics, we do have some ideas that might account for it. The best is that gravity 'leaks' into or out of our universe.

Leaks from another world

Various branches of modern physics suggest that there are many more dimensions of space than the three (up and down, side to side and backward and forward) that we are familiar with. One of the consequences of this is that certain forces can become 'diluted' through spreading into these extra dimensions. If the gravitational force is weak, that may be because it is spread more thinly than the others.

The 'extra' dimensions are thought to be 'compactified' – rolled up, essentially – so small we don't experience them in day-to-day life. It's just a theory at the moment, but a few researchers are trying to find evidence for this. One route is through examinations of the way the gravitational attraction between two objects changes with the distance between them.

Newton showed that gravity follows an 'inverse square law'. That means that the gravitational force one object exerts on another decreases in proportion to the square of the distance between them. Separate two objects by a metre, and measure the gravitational force. Then separate them by a further two metres and measure the attraction again. It will be nine times weaker because they are three times further away.

The hidden dimensions enter our world at submillimetre scales. If gravity behaves differently from normal on these very small scales – if the inverse square law doesn't hold when masses are separated by only a few thousandths of a milli-metre – that may be because these dimensions are interfering with things. Spot some disturbance here, then, and we might have evidence to support our most daring theories.

This is why physicists are carrying out the most exquisite experiments to probe gravity at microscopic scales. So far, however, they have found no evidence of violations of the inverse square law. That's a great shame, because one of the roles of these advanced, multidimensional theories is to improve our best theory of gravity, Einstein's relativity.

Gravity is relative

Einstein's theory of relativity cast space and time as a four-dimensional fabric and said the presence of mass or energy distorted this fabric. Where Newton had declared that bodies in motion will move in a straight line unless acted on by a force, Einstein added a twist. Yes, they moved in a straight line through space, but would have to follow any distortions in that space.

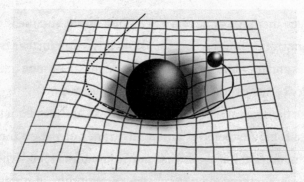

Gravity is a distortion of space–time

The distortion that the sun's mass creates, for instance, means that a nearby planet in motion will be pulled into a curved trajectory. Balance the masses and the speed of motion, and you have an orbit. Hence, in Einstein's view, gravity is a kind of illusion. Though it looks like a force that acts across space and time, it is actually more like topographical features – hills and valleys – added to the landscape, features that make it hard to travel in certain directions, and easier to travel in others.

Neat though this is, and supported by numerous experimental findings, we know it is not the final answer. In a way, Einstein has only given us a clever description of *how* gravity works. The *why* is still wide open. There is hope, though. Relativity, in its current form, is not compatible with quantum theory. We will have to wait for some future 'quantum gravity' theory to unite the two. And that theory, presumably, will give us the *why* of gravity, just as we have recently got to grips with the *why* of mass.

So far, we have been blithely talking about mass, while avoiding the obvious question. What does it mean that something has mass?

Physicists categorize mass in two distinct ways. One is 'gravitational mass', which is what produces and responds to gravitational fields. This is what makes the apple fall. The other is 'inertial mass', which is a measure of how hard it is to move something out of its current state of motion or rest. When you try to push a broken-down car, its inertial mass stands against you.

As far as we know, inertial and gravitational mass are entirely equivalent. Imagine standing on Earth in a sealed box like a stationary elevator. You feel the push of the floor as your gravitational mass responds to the influence of gravity. Now imagine taking that elevator box into space, away from gravitational fields, and sticking a rocket engine on it that accelerates it at 9.81 metres per second per second, the acceleration due to gravity at the Earth's surface.

There will be no difference in what you feel, Einstein says. This 'equivalence principle', which says there is no distinction between your gravitational mass and your inertial mass, is part of the bedrock of Einstein's general relativity. Although we don't have a definitive proof that it is absolutely correct, experiments have shown this is certainly true to at least one part in 10^{12}. A decade before he created general relativity, though, Einstein asked another question about mass. In 1905, his 'miracle year' when he also published his

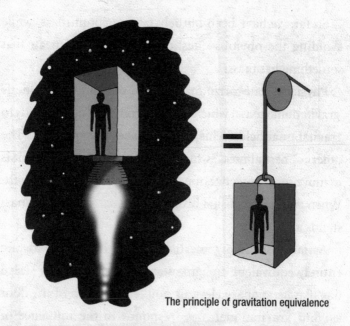

The principle of gravitation equivalence

special theory of relativity, Einstein came up with an interesting line of enquiry. He asked, in a landmark paper, whether the inertia of a body depends upon its energy content.

The energy of weight

This was the origin of the world's most famous equation: $E = mc^2$ (see *Why does $E = mc^2$?*). Energy and mass, in Einstein's view, could be interchanged. It took almost a century, but we now know through this equation that energy is indeed the root of mass. Take that apple, for instance. Its mass resides in its constituent components. Going down in scale, these are molecules, which are composed of atoms, which are composed of electrons, protons and neutrons.

The origin of the mass of the electron (which is only one-thousandth the mass of the proton and neutron) remains a mystery. But physicists are at least getting to grips with the mass of protons and neutrons. These particles are each composed of three particles called quarks. However, the masses of the quarks account for only around 1 per cent of the proton or neutron mass. The rest comes from a shadowy, quantum world of energy-stealing 'virtual particles'.

Down at the quantum scale, the rules are very different from those that we encounter in our day-to-day lives. Here, a phenomenon called the 'Heisenberg uncertainty principle' holds sway, and issues strange declarations. One is that nothing has a definite amount of energy, even when that energy is zero. Instead the energy fluctuates around zero, allowing seemingly empty space – something physicists call the 'vacuum' – to fizz with appearing and disappearing particles.

These particles appear in pairs: a particle and its antiparticle, spontaneously created as the energy of the vacuum of empty space fluctuates around zero. According to a Nobel Prize-winning branch of physics called 'quantum chromodynamics' (QCD), the particles can appear with various amounts of energy, giving a spectrum of characteristics. Sometimes they take a form where they are known to physicists as 'gluons'. Gluons create a force known as the strong nuclear force, which holds quarks together to create a proton or a neutron. And it is gluons – or rather their energy – that give the apple most of its mass. Working out exactly how much mass comes from the energy of all these virtual particles has not been easy, involving crunching combinations of around 10,000 trillion numbers. When the results came out, though, they were within a couple of per cent of the experimentally recorded masses of these particles.

The energy associated with the gluons, converted via Einstein's $E = mc^2$ formula, accounts for almost all the mass in a proton or neutron. There is a little missing: the mysterious electron mass, and a contribution from some more virtual particles, such as pairs of virtual quarks and antiquarks, and the Higgs boson (see *What is the God Particle?*). Essentially, though, the mass of the apple – and of the Earth – is a manifestation of the energy contained in the vacuum of empty space.

The success of quantum chromodynamics in establishing the origin of mass has given physicists hope that similar ideas will eventually lead us to the final *why* of gravity: the graviton. The electric and magnetic forces are manifest through atoms exchanging packets of energy called photons. The strong nuclear force comes via gluons, as we have seen. The weak nuclear force is known to result from the exchange of energy-laden particles known as the W and Z bosons. All of these have been seen in experiments. Gravity is thought to rely on the exchange of particles known as 'gravitons'. These, however, remain hypothetical. Despite all our advances in understanding, we still haven't seen a graviton.

That is not our only remaining problem with gravity, however – a much more embarrassing and basic issue remains unsolved. Bizarre as it may seem, although we have worked out the origin of mass using the most ingenious minds, the biggest computers and the greatest theories of physics, we still don't have a good way to measure the very thing that gravity acts upon: mass. Every other standard unit of measurement has a precise, atomic foundation. The second is based on a certain number of oscillations of a caesium atom. The metre is the distance light travels in a particular fraction of that second. The kilogram, though, is the mass of a lump of metal kept locked inside a Paris vault.

The changing kilo

It's not any old vault, of course: it is contained within the hallowed walls of the International Bureau of Weights and Measures (BIPM) near Paris. And it's not any old metal, either: it is a cylinder of platinum, chosen as the most stable, incorruptible material available. The mass of this platinum cylinder is the kilogram against which all other kilograms are calibrated. The problem is, its mass is changing. Metrologists have made dozens of copies, and the original no longer weighs the same. There is about 100 micrograms of difference, roughly equivalent to the mass of a couple of grains of salt. Researchers are planning ways to bring the kilogram into line with other standards, by using atomic measurements. One hope is to create a polished sphere of silicon containing a determinable number of atoms. The kilogram will then be defined as the mass of a certain number of silicon atoms.

Another possibility is to use something called a Watt balance to measure mass in terms of energy. Einstein told us that mass and energy are interchangeable; the Watt balance would invoke this by measuring mass against the energy contained in a carefully configured electromagnetic field. Until these plans come to fruition, though, we are stuck with plugging slightly inaccurate numbers into Newton's formula.

Gravity is everything to us – it pulled particles together to create the Earth, it holds us in orbit around our life-giving sun, it creates the tides that allowed that life to form and move on to land. And now we return the favour and use our gravity-given minds to make extraordinary discoveries about the very nature of this attraction. At the same time, though, we have only primitive tools to get the measure of it. While we can talk about gluons inside the nuclear structure of the atoms within the apple, we cannot be precise about how much the apple weighs. The essence of gravity remains deliciously difficult to tame.

Are solids really solid?
Atoms, quarks and solids that slip through your fingers

If all the world was made of gas, we could not exist. The way our bodies are organized, the way information is stored in the structure of DNA, the way our brains process and hold information, all requires that atoms are fixed in place, not floating freely around. Life, at least life as we know it, requires solidity. But what is a solid?

A gas is a collection of atoms or molecules that have no or extremely weak bonds between them. A liquid has weak bonds between the particles, allowing them to slip over and past each other. A solid, though, has its particles held together by strong electrostatic bonds. But that doesn't make a solid solid. Hold your hand up in front of your face. It looks solid enough, doesn't it? But to neutrinos, the tiny subatomic particles that flood the universe, your body is far from solid. Every second, trillions of neutrinos pass right through you, failing to interact with a single atom of your body. Scientific progress has made it clear that most of our solid matter is empty. We have even devised solid materials with the ghostly

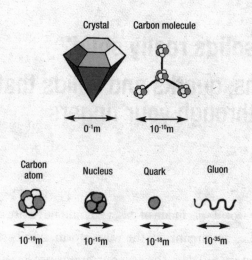

The diamond, and the elements that make it up

power to pass through each other. Experimental science is teaching us that the concept of 'solid' is a slippery one at best.

Our brains, another mass of solid atomic matter, have been able to probe this at a level even deeper than our experiments. Though there is no certainty here yet, our best understanding leads us to a remarkable conclusion: that there is no such thing as a solid. Every piece of matter is, essentially, the result of a random fluctuation in the energy of space and time. Solidity is, at its root, an illusion.

To explore this, let's start with a familiar solid. Something dependable, something robust. Diamond seems a good solid to test. It is the hardest naturally occurring material, and used as a tool to cut through the toughest of metals. How solid is diamond? It is diamond's molecular structure that makes it particularly hard. Its carbon atoms are bonded in a rigid tetrahedral arrangement, sitting about 10^{-10} metres apart from each other. Since it is the outermost electrons in the atom that form these bonds, it will come as no surprise to hear that this is roughly the size of the atom. But that doesn't make it truly solid. It is time for us to explore the strange world of atomic structure.

The first scientist to look into this question is generally considered to be Democritus. He was actually a Greek philosopher rather than a scientist, but he made a scientific conjecture about the nature of matter. All matter, he suggested, can be split so far, but no further. At the most fundamental level was the concept of *atomos*, from which we get our word atom. In Democritus's view, *atomos* were the particles that could not be split, destroyed or changed in any way.

And, until the earliest moments of the Industrial Revolution, that was essentially that. The age of the telescope came, and we learned to probe the heavens, but we made no progress in getting to the root of matter. That's because we needed tools that could influence matter on the atomic scale.

Inside the atom

It was the English schoolteacher John Dalton who kicked off the investigation of the atom. Towards the end of the 18th century, Dalton proposed that any single element was an assembly of identical atoms. These all had the same properties. Chemical reactions, he suggested, joined two different kinds of atoms together to form a chemical molecule. Dalton backed up his ideas with chemical experiments that determined the ratio of elements within certain substances, such as carbon dioxide: one part carbon to two parts oxygen.

The concept of atoms lent itself to the processes of the Industrial Revolution, enabling the pioneers of thermodynamics to work out the gas pressures and heat transfer rates that powered the rise of the machine. But we were no wiser about whether it might be possible to get inside an atom. In the age of the British Empire, the steam train and massive industrialization, the science of the atom had hardly moved on from the Greek idea of an indivisible substance.

Three near-simultaneous developments changed that. The investigations of English physicist Joseph J. Thomson revealed the existence of particles, which he called 'corpuscles', that were negatively charged and were 2,000 times lighter than even the lightest of the atoms. With this discovery, we had at last found something – we now call it the electron – that was smaller than an atom.

By 1904 Thomson was suggesting that atoms were composed of positive and negative parts mixed together to give a 'plum pudding' kind of structure. Around the same time, in Paris, Pierre and Marie Curie and Henri Becquerel discovered radioactivity. Their subsequent investigations showed that at least some of the activity resulted from the emission of charged particles from atoms. Back in England, meanwhile, the brash New Zealander Ernest Rutherford had arrived. In just a few decades of research, Rutherford was to make the greatest inroads into the atom for thousands of years.

Perhaps the most significant discovery was the revelation that Thomson's 'plum pudding' model of the atom was entirely wrong. Rutherford fired positively-charged alpha radiation particles – helium atoms stripped of their electrons – at a thin piece of gold foil. Almost all of the alpha particles passed straight through. Some, however, were strongly deflected. A few even bounced back at the emitter. This shocked Rutherford. 'It was as if you had fired a fifteen-inch shell at a piece of tissue paper and it had come back and hit you,' he later wrote.

Nuclear bombshell

To Rutherford, there was only one interpretation of this extraordinary result. A few of the positively charged helium atoms had happened to be fired directly at a concentration of positive charge, and been strongly repelled. Most of the volume of the atom was empty space. But at the centre lay all the positive charge – and almost all the mass. Rutherford had discovered the atomic nucleus.

The emptiness of an atom is hard to grasp, and provides our first clue to the illusion of solidity. The nucleus in the atom, Rutherford said, is 'like a gnat in the Albert Hall'. Others around him called it 'the fly in the cathedral'. Either way, it's a monstrous emptiness. If the nucleus was the size of a small apple, the edge of the atom, defined by the outer orbit of its negatively charged electrons, would be 3 kilometres (2.5 miles) in diameter. Each electron, meanwhile, would be smaller than the full stop at the end of this sentence. We can look at the emptiness another way. If you could remove the empty space in atoms, and pack hydrogen nuclei into the volume of a penny with no space between them, you would have a penny-sized object that weighed more than 30 million tons.

Inside the nucleus

As the lightest element, hydrogen has the simplest possible nucleus: a single positive nuclear charge, or proton. But, generally, there is more to nuclei than just the proton. The carbon atoms we have been examining, for example, have a much more complex nucleus, containing half a dozen uncharged particles called neutrons. All atoms (apart from hydrogen) contain neutrons. The neutron, which is very slightly heavier than the proton, was discovered by James Chadwick at the University of Liverpool in the early 1930s. Carbon has a nucleus composed of six protons and,

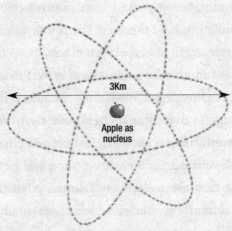

3Km

Apple as
nucleus

The emptiness of the atom

depending upon the exact 'isotope' we are dealing with, six, seven or eight neutrons.

So, is there any solidity here? Rutherford found the proton to be around 10^{-15} metres in diameter. The neutron is of approximately the same size. And atomic nuclei don't mirror the emptiness of the atom. The carbon nucleus is no bigger than one would expect if the particles within it were packed tightly together. Larger nuclei make the tight packing of the nucleus even clearer. A uranium nucleus, which contains 238 particles, is only 14 proton-widths in diameter – it is rather like a basketball stuffed with 238 ping-pong balls.

With this discovery, physicists had a notion of solidity at the core of matter. But only for a while: things soon got slippery again, taking us on a downward spiral that, today, tells us there is nothing solid in the entire universe. The problem is that, being packed with positive charges, the nucleus should not hold together. The protons in a carbon nucleus should, by rights, repel each other.

That means another force must be at work. Physicists call this the 'strong' nuclear force simply because it has to be strong enough to overcome the repulsive electromagnetic force. To investigate the strong force required physicists to delve into the characteristics of the proton and neutron, or nucleon, as they are collectively known. What they discovered was that the nucleons were not fundamental, indivisible particles, but composed of three 'quarks'.

Quarks at heart

The name 'quark' was chosen by physicist Murray Gell-Mann in 1964, who picked it out after reading the phrase 'Three quarks for Muster Mark' in James Joyce's *Finnegan's Wake*. The quark started out as a hypothetical particle, whose existence was also suggested, independently, by the Russian-American physicist George Zweig (who wanted to call it an 'ace'). Both men's guess turned out to be a good one – though it took a good while to prove it.

Physicists can only see matter on this scale by smashing together subatomic particles in accelerators. The collisions create smaller particles, whose fleeting existence has to be inferred by the trails left behind in detectors that line the walls of the accelerator at the collision site. The first quarks were identified from collisions at the Stanford Linear Accelerator Centre (SLAC) in 1968. Two more decades passed before all the hypothesized quark particles had been seen. But we now know that quarks come in six 'flavours', exotically named: strange, charm, top, bottom, and the much more common up and down.

Protons are composed of two up quarks and one down quark; neutrons are two down quarks and one up quark. But it is the top quark that may be the undoing of solidity. The top quark is unaccountably heavy. It weighs almost the same as a gold atom, which is why it took until 1995 for our particle accelerators to be able to make one. Particle accelerators are governed by $E = mc^2$, and it takes a great deal of energy to make so much mass.

A gold atom contains 79 protons and 118 neutrons. That is a total of nearly 600 up and down quarks. How can just one top quark weigh nearly the same? Something in the nature of quarks, and how they come together, suggests there is a mystery in the nature of mass. A theory called quantum chromodynamics (QCD) makes that clear. It has shown that the up and down quarks that make up protons and neutrons account for only 1 per cent of the mass of these particles. The rest is, as provided by $E = mc^2$, held in the energy that binds the quarks together. This is the 'strong' nuclear force.

Sensing the energy of emptiness

According to QCD, the strong force has its roots in the uncertainty principle of quantum mechanics (see *Is Everything Ultimately Random?*). This principle says that nothing that can be measured actually has a precisely defined value. That even applies to empty space: it can't have exactly zero energy. As a result, empty space has a fluctuating but finite amount of energy.

This fluctuating energy manifests as particles called gluons, and it is gluons that create the strong force that binds the quarks. So when you hold a diamond in your hand, you feel its weight. But what you sense as the mass of the diamond is actually the result of a shifting, shimmering energy field that creates the weight of the quarks that make up the protons and neutrons in the nucleus of each carbon atom. In a sense, that diamond, that most solid of objects, doesn't have a permanent existence at all. As it rests on your hand, all that is happening is that a continuum of energy fluctuations are manifesting as solidity.

Slippery solids

Perhaps we shouldn't be surprised that the rules of solidity are turning out to be flexible. Solids, after all, are only solid under certain conditions. Heat up an ice cube, and it will create a pool of water. The molecules haven't changed their essential nature; it is simply that the environmental conditions have altered the strength of the bonds between them. The same is true when we heat the water and it turns into steam. Now the bonds between the molecules have disappeared – but still the molecules themselves haven't changed.

We can create a new kind of matter, at the other end of the temperature scale, too. As we cool some kinds of materials down, we can create a new kind of matter. To solid, liquid and gas, we can add the phase known as the 'Bose-Einstein condensate'. The BEC is the result of a radical transformation that only happens at extremely low temperatures. Temperature is, in essence, a measure of how much energy an object has to 'jiggle about'. At very low temperatures, a material is stripped of all energy, and so hardly moves at all. But quantum theory dictates that the more precisely you pin down an object's momentum – in this case to near-zero – the bigger the uncertainty in its position. So every particle in the BEC has an uncertain position. In effect, all the particles overlap each other, merging into one big quantum object, like a giant atom.

In this state, all kinds of strange behaviours arise. When niobium metal turns into a BEC, the quantum laws turn it into a 'superconductor' that carries electrical current without any of the resistance associated with currents in normal metals. When helium atoms form a BEC, for example, a similar thing happens: stir a cup of this 'superfluid' helium, and the swirl goes on swirling for ever. Even more bizarrely, superfluid helium can defy gravity, flowing up the sides of a container. Turn helium into a solid, where its atoms are held together in a crystal, and the weirdness gets worse.

Not that it's easy to make solid helium. To get it to a liquid requires cooling it to within 4 degrees of absolute zero. To turn that liquid into a solid requires crushing the atoms together: the liquid has to be cooled to within 1 degree of absolute zero and compressed with 25 times normal atmospheric pressure. Once you're there, though, you can see the strangest solid in the universe.

The bonds between the atoms in solid helium are extremely weak. So weak, in fact, that atoms can break off. This leaves what is known as a 'vacancy' in the crystal. Physicists have long known that these vacancies can be treated like particles in their own right. They are really like an atom with slightly different properties. They affect the way a material conducts electricity, for example; it is only because of vacancies that semiconductors have the properties they do. The entire multi-billion dollar business of electronics relies on the properties of vacancies.

In an ultra-cold helium crystal, the laws of quantum mechanics lock all the vacancies in the structure together to form a vacancy-based BEC. With the atoms locked together too, the helium crystal becomes two 'supersolids'. And, if you get the experimental conditions right, they can pass right through each other. In theory, any solid crystal will behave in this way under the right conditions.

It might not even require the formation of vacancies: in some materials it ought to be possible to make all the freed atoms lock together and move around the crystal as one, meaning that the solid will pass through itself. It's not unlike the strange conjuring tricks where two solid rings are made to pass through each other, lock together and then, with a flourish of the magician's hand, come apart again. In this case, though, it is the solidity that is the illusion.

Look at your hand again. It is mostly made of nothing. The crystal structures of the proteins leave enormous gaps between the tiny atoms. The atoms themselves are almost entirely devoid of matter. Where there is matter – in the atomic nucleus – most of its mass is derived from quantum fluctuations in the energy of empty space. The solidity of that hand in front of your face is perhaps the most convincing illusion you will ever experience.

Why is there no such thing as a free lunch?

Energy, entropy and the search for perpetual motion

The exact origins of the phrase 'no such thing as a free lunch' are unclear, but most sources say it began life as the pithiest summary of economics. It appeared in Pierre Dos Utt's 1949 monograph *TANSTAAFL: a Plan for a New Economic World Order*, where Dos Utt tells of a king seeking economic advice. His advisers, looking for ever-simpler ways to get their message across, conclude with the now-classic version of the phrase: 'There ain't no such thing as a free lunch.'

It is doubtful this would have been enough to motivate economists to usher in a new world order, and the physicists of the time would have certainly been unimpressed. The idea of something for nothing had long been a goal of inventors trying to get a free lunch by coming up with 'perpetual motion machines' that would do work without the need for any external power. Physicists had long been telling them this was impossible.

There is no such thing as a free lunch because you simply can't get something for nothing: someone, somewhere always has to pay. Physicists have enshrined this principle as a fundamental law of physics. So you need to think hard before you start looking for a free lunch, because you are battling against the way the universe runs. Perhaps the great artist, visionary and inventor Leonardo da Vinci put it best. He took a keen interest in perpetual motion, investigating designs, and coming up with a few of his own. But he was sceptical about them all: one of his notebooks contains a detailed analysis of a popular kind of machine, showing why and how it could not work. 'O you researchers of perpetual motion,' Leonardo wrote, 'how many harebrained ideas have you created in this search. You may as well join the alchemists.'

There are two kinds of perpetual motion machines. The first supplies an endless output of work despite the fact that there is no input of fuel or any other form of energy. The second converts heat to mechanical work with perfect efficiency. Both, it should be made clear, are wishful thinking – and physics tells us why.

Something for nothing

As with alchemy, the search for perpetual motion engaged some of the finest minds that have graced the Earth. The dream has been around since at least AD 624, when the Indian mathematician and astronomer Brahmagupta described a wheel whose hollow spokes could be filled with mercury. The mercury would shift weight around the wheel as it rotated. As a result, Brahmagupta wrote, 'the wheel rotates automatically for ever'.

The idea was repeated numerous times. In 1235, Villard de Honnecourt, a French artist and inventor, produced his own version. De Honnecourt was no fool: he drew the first known plans for a mechanical escapement mechanism that would keep time. But de Honnecourt's 'overbalanced wheel' still doesn't work. Here, a series of hinged weights are attached around the circumference of a wheel, their motion limited by pins. As the wheel turns, an imbalance in the distribution of weights causes the wheel to turn. As it turns, the elevated weights drop onto their pins, and the transfer of weight keeps the wheel turning.

The fact that the perpetually rotating wheel is a running theme in the search for perpetual motion can only mean that very few people tried to build these kinds of machines. Build one and you soon learn that they just don't work. Take de Honnecourt's overbalanced wheel, for example. What is needed for this to carry on forever is for the uppermost rod to flip over as it reaches the top of the wheel, maintaining the imbalance. Unfortunately, this doesn't happen: the weight distribution is such that it doesn't quite flip. After one revolution, the weights return to their initial position, and everything is back exactly where it started – including the stationary wheel.

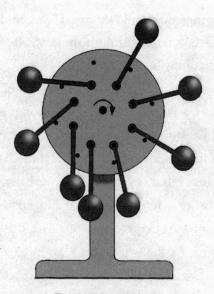

The overbalanced wheel

To be fair to de Honnecourt, the reason for this was not clear until well after his time. The problem is that energy is transformed between two different forms. Because the rods have the potential to fall under the influence of gravity, they are said to have 'potential energy'. If the wheel turns, some of this converts to the 'kinetic energy' of movement. However, after one cycle, the rods return to their initial position, and therefore must have exactly the same potential energy (which is due to their position) as before. Since there is no external source of energy, and the rods have the same potential energy at every turn, there is nothing to put energy into turning the wheel.

Energy is conserved

By 1775, the Royal Academy of Sciences in Paris had had enough of perpetual motion. It issued a statement declaring that the Academy 'will no longer accept or deal with proposals concerning perpetual motion'. And in 1841, scientists finally found a scientific principle to throw at perpetual motion seekers: the first law of thermodynamics.

It was the first explicit statement of the conservation of energy. Leonardo da Vinci had suggested that, 'Falling water lifts the same amount of water, if we take the force of the impact into account,' but it took the German physicist Julius Robert Von Mayer to explore the matter properly and issue an edict. Energy, he said, cannot be created or destroyed.

Not that he was taken seriously straight away: Von Mayer was told, for instance, to find some experimental evidence to back up this strange idea. This he did, by showing that the kinetic energy of vibration could be transferred to water molecules, manifesting as an increase in temperature. Once the point was proven, the principle was quickly accepted by physicists, and used to keep perpetual motion at bay. Motion takes energy, and the conservation of energy principle tells us that you can't get more energy out of a closed system than is there in the first place. Since friction affects any and every mechanism, dissipating some of that energy as heat and sound, inventing perpetual motion machines of the first kind became a fool's errand. Not that this put the perpetual motion seekers off. Around this time, the science of thermo-dynamics was giving them a whole new lease of life. Their goal? Perpetual motion machines of the second kind.

Miracle machines

The second kind of perpetual motion machine is something that extracts heat energy from a reservoir, such as the air or the ocean, and converts it into mechanical energy. It certainly seems like a good idea. The oceans are so vast a resource that, if we could extract heat that would cause a one degree drop in ocean temperatures, it would supply something like the energy needs of the United States for half a century.

The plausibility of this kind of machine is enticing. Indeed, creating an efficient steam-powered engine has been a human obsession since Hero of Alexandria created the 'aeolipile' in AD 1. This ball, that was set rotating by jets of steam, had no particular uses. However, subsequent inventions used steam turbines to turn spits, pump water from mines and power grinding pestles. None of them got anywhere near a truly useful efficiency, however. That efficiency came with James Watt's steam engine, first demonstrated in 1765. It was a development of the engine invented by Thomas Newcomen, and raised the efficiency enough to kick off the Industrial Revolution. The theory behind such engines, though, was still very much in development. The builders of steam engines were working on hunch and intuition, not scientific theory.

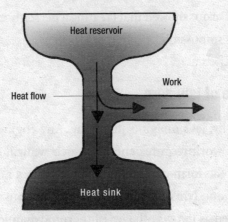

The Carnot engine

It wasn't until 1824 that the French military scientist Sadi Carnot published *On the Motive Power of Fire*. Even then, this primary work in the field went largely unnoticed for a decade. But the scientific principles behind the steam engine were now in place. And, as a bonus, Carnot had worked out the principle that denies a free lunch to perpetual motion machines of the second kind.

There is a good reason why you can't get useful work out of a room temperature heat source. It is called the second law of thermodynamics, and it says, essentially, that you can't take the heat from something then turn all the heat into mechanical work. Some of that heat has to be passed on to a 'heat sink' at a lower temperature. It is the temperature difference between the heat source and the heat sink that determines how much work you'll get out of this 'heat engine'. Carnot showed that creating a perfectly efficient heat engine is impossible.

The rule of law zero

To see why, let's imagine an engine. Any engine seeking to perform work requires energy, which we will consider to come in the form of heat. Heat flows from a hot source to a colder one (this principle seems so obvious it was only formalized as the 'zeroth law' of thermodynamics long after the other laws were laid down), so both reservoirs are required; work can be extracted as heat flows from a hot 'reservoir' to a cold one.

The work extracted in this situation is the difference between the heat flowing out of the hot reservoir and the heat flowing into the cold reservoir. A perfect efficiency would have zero heat flowing into the cold reservoir so that all of the heat energy is used for the work you want to do.

Now let's consider, as Carnot did, the practicalities of the engine. Carnot imagined a piston engine much like the cylinder of a car engine, where the heat is used to expand gas that pushes on a piston. The gas is then compressed, and the cycle begins again. By considering the gas laws that relate pressure, temperature and volume, Carnot showed that the efficiency of an engine depends upon the ratio of the temperatures of the hot and cold reservoirs. No matter what fluid or gas is being used to power the engine, the ratio of the two temperatures is everything. And here is the problem with this free lunch.

The average diesel engine operates at around 550 celsius. The exhaust gases exit to the outside temperature. The maximum efficiency possible, according to Carnot's work, is around 60 per cent. In reality, a diesel-powered car converts around 50 per cent of its fuel's chemical energy into energy that can move the car along the road. The rest is wasted as heat (which is why cars need cooling systems). Petrol engines are significantly less efficient.

What if we operate the two reservoirs at the extremes of temperature? In theory, the hot reservoir can operate at infinitely high temperatures. But the cold reservoir cannot be colder than absolute zero. Even dumping the heat in outer space would give a cold reservoir temperature of 3 K, or −270 celsius. Because you can't get lower than zero, and an infinitely hot reservoir does not exist (at least not one that we know about), a perfectly efficient engine is impossible. You cannot convert heat into work without wasting some of the heat. And that means that, to continue the cycle, you always have to put in energy. No free lunch, in other words.

Carnot's work led directly to the formulation of the second law of thermodynamics. As phrased by the English physicist Lord Kelvin and the German physicist Max Planck, it states that an engine operating in a cycle cannot transform heat into work without some other effect on its environment. Thanks to the second law, not only can you not get a free lunch, you can't even keep your lunch cool in the refrigerator for free. Refrigeration, it turns out, is nothing more complicated than the Carnot engine working in reverse.

In 1850, the German physicist Rudolph Clausius rephrased the second law to read, 'An engine operating in a cycle cannot transfer heat from a cold reservoir to a hot reservoir without some other effect on its environment.' A refrigerator, in other words, needs energy put into it. This arises from the natural tendency of energy to flow 'downhill': from hot to cold. Keeping the inside of your refrigerator below the temperature of your kitchen involves the same process of expanding and contracting, heating and cooling gases as running your car engine, and it all takes energy. This time, though, you need a compressor rather than an expander for the gas.

The march of entropy

As mentioned, Carnot's work involved consideration of the pressure, temperature and volume of the gas. The process that Carnot uncovered led to another revelation for physicists: the notion of entropy. The whole universe, it turns out, is spiralling into ever-more disorder. It was Clausius who classified this disorder as 'entropy', a word derived from the Greek for 'transformation'. In 1865, he wrote a mathematical treatise on the work that the atoms do on one another in a gas. The result, Clausius showed, is that the second law can be expressed in a new way: the entropy, or disorder, of a closed system either stays the same or increases – it never decreases.

That doesn't mean you'll never see entropy decrease on a small scale. Your lunch inside the refrigerator will get cold, for instance, decreasing the disorder in its constituent molecules. But don't be fooled that this breaks the second law of thermodynamics. The inside of your fridge is not a closed system – the molecules of refrigerant gas take the heat away, and their disorder increases as they do. As the heat is transferred to the air in your kitchen, the disorder in your house increases too.

This kind of thing is happening throughout the universe as the processes of nature unfold. It creates, in physicists' view, the irreversibility of natural processes: the arrow of time is just another way of expressing the second law of thermodynamics. The wasted energy of Carnot's engine cycle is the slow unravelling of the universe in microcosm.

It's worth pointing out that, in some physicists' minds, we are only here because of a free lunch: the universe. According to Alan Guth, the universe is the ultimate free lunch. Guth is the originator of an idea in cosmology called 'inflation'. According to Guth, the universe, and all the energy it contains, seems to have arisen from little more than a gram of material. A fraction of a second after the Big Bang, the universe was a 100 billion times smaller than a proton, but it then blew up like a balloon. In fact, it blew up like a pea expanding to the size of the Milky Way in less time than it takes to blink an eye.

The numbers involved are staggering. It started when the universe was around one billionth the size of a proton. 10^{-34} seconds later it had expanded to 10^{25} times its original size – something around the size of a marble. And during this process, cosmologists reckon the energy within the universe increased by a factor of 10^{75}. It sounds like a violation of the something from nothing, or no free lunch, rule. But there's a complication that keeps it within the laws of physics: some of it is negative energy.

According to general relativity, our best description of the nature of space and time, the energy of a gravitational field is always negative. During inflation, the energy in the rapidly expanding space–time becomes ever more negative. Within this space–time, however, matter began to appear. That's because the properties of space–time mean that a portion of it spontaneously moves to a lower energy state: particles such as electrons, positrons and neutrons. Matter has positive energy, and the continuing creation of matter created more and more positive energy to balance the growing negative energy. The total energy can thus remain constant. The ancient Greeks said that nothing can be created from nothing, but inflation begs to differ.

However, in today's universe, the first and second laws of thermodynamics put up a brick wall to any claims for the generation of a free lunch. So well proven are they, in fact, that the US Patent Office warns anyone submitting a patent for a perpetual motion machine that they should think carefully; they will most likely lose their money. 'The views of the Patent Office are in accord with those scientists who have investigated the subject and are to the effect that such devices are physical impossibilities,' the office's official statement says. 'The position of the Office can only be rebutted by a working model . . . The Office hesitates to accept fees from applicants who believe they have discovered Perpetual Motion, and deems it only fair to give such applicants a word of warning that fees cannot be recovered after the case has been considered by the Examiner.' So not only is there no such thing as a free lunch; even looking for one could end up costing you money.

Is everything ultimately random?

Uncertainty, quantum reality and the probable role of statistics

We could start by turning the question on its head. Is everything predictable? Can we work out the rules that determine how the processes of the universe occur? That would give us extraordinary power over nature, the kind of power humanity has always dreamed of.

In many ways, the whole of human existence is wrapped up in this quest. We look at the world around us, and attempt to find regularities and correlations that enable us to reduce what we see to a set of rules or generalities. This enables us to make predictions about the things we might or might not encounter in future, and to adjust our expectations and our movements accordingly. We are, at heart, pattern-seekers.

A facility for pattern-spotting has served us well as a species. It is undoubtedly what enabled us to survive in the savannah. A predator might be camouflaged when still, but as soon as the beast moved, we would spot a change in the patterns in our surroundings, and take evasive action. Roots and berries grow in predictable geographical and temporal patterns (the seasons), enabling us to find and feast on them.

The evidence suggests that, because our lives depended on pattern recognition, the evolution of our brains took the process to extremes, forcing us to see patterns even when they are not there. For example, we over-interpreted the rustles of leaves and bushes as evidence for a world of invisible spirits. Modern research suggests this kind of over-sensitivity to patterns in our environment has predisposed us to religious conviction; a tendency towards irrational thinking – the consideration of things we can't touch, see or account for – is the price the human species has paid for its survival.

Ironically, though, scientists have only been able to draw conclusions about where irrational thinking comes from because of the mote in their own eye. Scientists are now painfully aware of their tendency to see patterns where there are none, and randomness where there is order. In order to combat this, and to determine whether there is any order, purpose or structure in the world around us, we needed the invention that exposes both how brilliant, and how foolish, the human mind can be. You might know it better as statistics.

The die is cast

Unlike many of the developments of modern science, statistics had nothing to do with the Greeks. That is remarkable when you consider how much they enjoyed gambling. The Greeks and Romans spent many hours throwing the ancient world's dice. These were made from *astralagi*, the small six-sided bones found in the heels of sheep and deer. Four of the sides were flat, and these were assigned the numbers. Craftsmen carved the numbers one and six into two opposing faces, and three and four into the other two flat faces. The way the numbers were situated made one and six around four times less likely to be thrown than three or four.

An enterprising Greek mathematician, you might think, could have made a fortune in dice games involving *astralagi*. However, there are reasons why no one did. Firstly the Greeks saw nothing as random chance: everything was in the hidden plans of the gods. Also the Greeks were actually rather clumsy with numbers. Greek mathematics was all to do with shape: they excelled at geometry. Dealing with randomness, however, involves arithmetic and algebra, and here the Greeks had limited abilities.

The invention of algebra was not the only breakthrough required for getting a grip on randomness. Apparently, it also needed the production of 'fair' dice that had an equal probability of landing on any of their six faces; the first probability theorems, which appeared in the 17th century alongside Newton's celestial mechanics, were almost exclusively concerned with what happens when you roll dice.

These theorems were predicated on the idea that the dice are fair and, though rather primitive, they laid the foundations for the first attempt to get a handle on whether processes in the natural world could be random. From dice, through coin tosses and card shufflings, we finally got to statistics, probability and the notion of randomness.

Probable cause

The Belgian astronomer and mathematician Adolphe Quetelet first began to apply probability to human affairs in the 1830s, developing statistical notions of probability distributions of the physical and moral characteristics of human populations. It was Quetelet who invented the concept of the 'average man'.

When he turned his attention to the notion of randomness in natural events, Quetelet was determined to take no prisoners. 'Chance, that mysterious, much abused word,

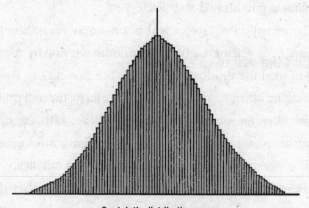

Quetelet's distribution curve

should be considered only a veil for our ignorance,' he said. 'It is a phantom which exercises the most absolute empire over the common mind, accustomed to consider events only as isolated, but which is reduced to naught before the philosopher, whose eye embraces a long series of events.'

Though ancient civilizations might have been able to predict the motions of the planets, until Quetelet no one thought that there could be any pattern to the way rain falls on a windowpane or the occurrence of meteor showers. Quetelet changed all that, revealing statistical patterns in things long thought to be random.

Not that the notion of randomness was killed with Quetelet. His work showed that the 'long series of events' followed a statistical pattern more often than not. But that left open the idea that a single event could not be predicted. While a series of coin tosses will give a predictable distribution of heads and tails, the outcome of a single coin toss remains unpredictable in Quetelet's science.

Lifting the veil of ignorance

Even here, though, science has now shown perception of randomness to be a result of ignorance. Tossing a coin involves a complicated mix of factors. There is the initial position of the coin, the angular and linear momentum the toss imparts, the distance the coin is allowed to fall, and the air resistance during its flight. If you know all these to a reasonable accuracy, you can predict exactly how the coin will land.

A coin toss is therefore not random at all. More random – but still not truly random – is the throw of a die. Here the same rules apply: in principle, if you know all of the initial conditions and the precise dynamics of the throw, you can calculate which face will end facing upwards. The problem here is the role of the die's sharp corners. When a die's corner hits the table, the result is chaotic (see *Does Chaos Theory Spell Disaster?*): the ensuing motion depends sensitively on the exact angle and velocity at which it hits. The result of any subsequent fall on a corner will ultimately depend even more sensitively on those initial conditions. So, while we could reasonably expect to compute the outcome of a coin toss from the pertinent information, our predictions of a die throw will be far less accurate. If the throw involves two or three chaotic collisions with the table, our predictions may turn out to be little better than random.

It is important to make the distinction between chaotic and truly random systems, however. A dice throw is not predictable to us, but neither is it random: we know it follows discernible laws, just not ones whose consequences we can accurately compute given our limited knowledge of the initial circumstances. We can say the same about the weather: it is our limitations – our ignorance, in Quetelet's words – that make it seem random. So is anything truly random? This is a question that lies at the centre of one of the greatest, and most fundamental, debates in science.

At the beginning of the 20th century, Lord Kelvin expressed his delight at the way physics was progressing. Newton had done the groundwork, and his laws of motion could be used to underpin the emerging understanding of the nature of light and heat. Yes, there were a couple of small issues – 'two clouds', as he put it – but essentially physicists were now doing little more than dotting the 'i's and crossing the 't's on our understanding of the universe. Coincidentally, the great German mathematician David Hilbert was feeling similarly optimistic. In 1900, at a mathematical congress in Paris, Hilbert set out 23 open mathematical problems that, when solved, would close the book of mathematics.

Certain about uncertainty

Both Hilbert and Lord Kelvin were guilty of misplaced optimism. Within a few years, relativity and quantum theory had blown apart the idea of using Newton to formulate the future of physics. What's more, the Austrian mathematician Kurt Gödel had pulled the rug from under Hilbert's feet, answering a mathematical question that Hilbert had not even asked – and taking away all certainty that any of Hilbert's questions could be answered.

Gödel had formulated what he called an incompleteness

theorem. It says, essentially, that there are some mathematical problems that can never be answered. Because of the way we formulate mathematical ideas, some things can never be proved. Mathematics is destined to be eternally incomplete. This has deep relevance for the question of randomness. If some things are unknowable, their behaviour may be, for all we know, random. Randomness might not actually be an inherent property of the system, but we can never prove that it is not. Gödel published his incompleteness theorem in 1931. By this time, the notion of limits to what we can know was hardly even a surprise. If you were familiar with the newly birthed quantum theory, you were already resigned to your ignorance of the ultimate answers.

First, quantum theory gave us the problem of inherent uncertainty. Werner Heisenberg was the first to notice that, when dealing with the equations of quantum theory, you could ask questions about the characteristics of the system under scrutiny, but there were certain combinations of questions that couldn't be asked simultaneously. The equations will give you the precise momentum or position of a particle, for instance. But they won't give you both at the same time. If you want to know the precise momentum of a particle at a particular moment, you can say literally nothing about the position of the particle at the same moment. This, the Heisenberg uncertainty principle, is a fundamental characteristic of quantum theory.

Heisenberg used the analogy of a microscope to justify

this. If we want to look at the position of a particle, he said, we have to bounce something off it – a photon of light, in this case. But by so doing the photon imparts momentum to the particle. In other words, by measuring the position, we have introduced a change to a separate characteristic – we cannot simultaneously know the position and the momentum with any accuracy. Any measurement – to determine momentum, or energy, or spin – will have concomitant effects on other characteristics. Certainty about every characteristic of a system at one moment in time can never be achieved.

The second even more fundamental problem is not so much one of practical limitations, but straightforward inherent predictability. The classic example for this is a piece of radioactive rock, such as the lump of radium that Marie

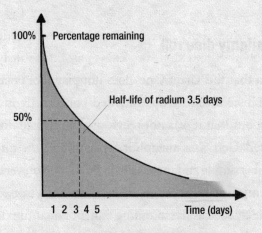

Radioactive decay of radium 224

Curie carried around with her. Physicists can tell you that, if it is composed of the quickest-decaying isotope of radium, the radioactivity of the lump will be halved every three and a half days. After a week, then, it has a quarter of its original radioactivity.

This, however, is a statistical average. It tells you nothing about whether any particular atom of radium will decay in a given time. After 1,000 years, some of the atoms in that lump will still not have decayed. Some will decay in minutes of you starting your clock. And there is no way to predict which is which. Nothing in quantum theory tells us what prompts the decay. It is, to all intents and purposes, random, as if the Almighty rolls ten dice for each atom and only a set of ten sixes causes decay. Einstein took this as proof that quantum theory is incomplete. There must, he said, be some 'hidden variables' that substitute for this divine dice game.

The Almighty dice roll

The idea that the 'Great One' does not play dice is perhaps Albert Einstein's most documented concern. It is worth pointing out that it was not religiously motivated. Einstein often used 'God' as a metaphor for nature or the universe. His point is simple and materialistic. Surely the universe runs by deterministic laws? Surely every effect has a cause? Niels Bohr, widely seen as the founding father of quantum theory, responded to this with scorn. Quantum theory, he told

Einstein repeatedly, is founded on randomness. Some effects have no cause. 'Einstein, stop telling God what to do,' he said.

As with Heisenberg's uncertainty principle, this randomness does seem to be written right into the equations of quantum theory. The central equation, the only way to make sense of experiments carried out on quantum systems, is the Schrödinger wave equation. This assigns quantum objects the characteristics of waves. If we want to know something about the quantum world, we solve this wave equation. All we get out of it, though, is a probability.

This is really what sets quantum theory apart. By the time quantum theory was born, in the 1920s, statistics was a firmly established discipline of science. Thermodynamics, the study of heat that had partnered the Industrial Revolution, relied upon it. Many other branches of science used statistics to verify the results of experiments. Quantum theory, though, seemed unique – and, to Einstein, disturbing – in its assertions that its results could *only* be expressed as probabilities.

The results of quantum experiments, according to the orthodox interpretation of quantum theory, were down to pure chance. Einstein's refusal to accept this is largely to do with the profundity of its implications. Quantum theory describes the world at the scale of its most fundamental particles. If quantum processes are random, then *everything* is ultimately random.

Bohr had no problem with this because he believed that, ultimately, nothing has any properties at all. Our experiments and measurements, he believed, will produce certain changes in our experimental equipment, and we interpret those in terms of the momentum of an atom, or the spin of an electron. But, ultimately, he said, those qualities are not a reflection of something that exists independently of the measurement. Thus, to Bohr, there was no reason why the results of experiments should not appear randomly distributed; there was no ordered objective reality from which some non-random result could arise. To his mind, it would be odd if it were any other way.

It seems an extraordinary point of view: radical and shocking. An electron only exists as some quirk of our measuring apparatus. Small wonder that the infinitely more 'common sense' oriented Einstein debated this with Bohr for decades. Einstein said he 'felt something like love' for Bohr when it began, such was the intensity and pleasure of their intellectual jousting. However, by the end, it had reached the point where the pair had nothing to say to one another. One dinner given in Einstein's honour saw Einstein and his friends huddled at one end of the hall, while Bohr and his admirers stood at the other.

Living with a random universe

Ultimately, history has decided that Bohr was right. Perhaps that is inevitable, given the force of Bohr's personality – he did once reduce Werner Heisenberg to tears, for instance. Whatever the truth, while Einstein's notion of a set of hidden variables that are waiting to be discovered remains scientifically respectable, the mainstream view is that objective reality does not have any independent existence. All we can say about the reality that manifests in quantum experiments is that we can predict the spectrum of possibilities, and how likely each one is to be seen. So, is that the last word? Is the universe ultimately random? Are we, as creatures composed of quantum molecules, doomed to find ourselves at the mercy of capricious forces? Yes – but the question is loaded like a Roman die.

We naturally seem to phrase randomness in negative terms, talking about suffering 'the slings and arrows of outrageous fortune'. But, as Shakespeare well knew, luck is often kind too. He has Pisanio declare in *Cymbeline*, for example, that 'fortune brings in some boats that are not steered'. The problem is, millennia of religious thought has imposed a sense that everything that happens in the world around us happens for a reason. Science has reinforced this: we appreciate predictability. But randomness can be useful too.

What's more, it may even be at the root of our very existence. The Heisenberg uncertainty principle is, as we have seen, fundamental to the universe. One of the consequences of this is that even regions of empty space cannot have zero energy; instead, all of space is populated by a frothing of 'virtual' particles that pop in and out of existence at random. These quantum fluctuations in the 'vacuum' of space are thought to be the source of the 'dark energy' that is driving the accelerating expansion of the universe. A similar kind of fluctuation that came 'out of nothing', but grew rather than disappeared again, is the best explanation we have for the cause of the Big Bang that gave rise to our universe. You might think randomness is a bad thing, but without it you wouldn't be here to think at all.

What is the God Particle?
The Higgs boson, the LHC and the search for the meaning of mass

You may not be surprised to learn that it has nothing to do with God. Except, perhaps, in the sense that no one has ever proved it exists. Nobel Prize-winning physicist Leon Lederman coined the phrase. Partly it was a humorous dig at physicists who thought that a particle would answer all their questions about the universe, partly it was a dig at the idea that science's discoveries might have anything to say about the meaning of life.

Unfortunately, the God particle does neither: it won't tell us everything about the universe, and it won't tell us the meaning of life. But that doesn't mean the Higgs boson is not worth searching for. It is the final piece of the puzzle in the standard theory of particle physics. If it exists, we can rest assured that we have uncovered much of the essential nature of the universe and found what gives materials their mass. If it doesn't, we might just have to go back to the drawing board.

The stage upon which this drama is playing out is in Switzerland. At the European Organisation for Nuclear

Research (CERN) in Geneva, the world's most powerful particle accelerator will be the arbiter of the truth in what physicists call the 'standard model' of physics. As the Large Hadron Collider (LHC) smashes together protons with the force of two high-speed trains colliding, the God particle may spill out. Around the world, physicists have been on tenterhooks, waiting to see if, way back in 1964, Peter Higgs was right.

Birth of the Higgs boson

Peter Higgs's suggestion was fairly straightforward. Responding to various attempts to work out the origin of mass, he wrote a paper that described how theoretical physics allowed for the existence of a new kind of field. It would be an addition to the known fields, such as the gravitational or electromagnetic field. This new field would have appeared as the universe cooled down from the Big Bang fireball, and could provide a source of drag on certain types of particles, perhaps endowing them with the property we know as mass.

The paper was initially rejected by the editors of the journal *Physics Letters* as being 'of no obvious relevance to physics'. Higgs rewrote it to give the idea a concrete application: it might arise, he said, in the force that holds the particles in the nucleus together, but still no one made much of it. Until, that is, Steven Weinberg, Sheldon Glashow and

Abdus Salaam set about trying to unify the electromagnetic and weak nuclear forces.

The theories of these two forces seemed uncannily similar in many aspects. The theory of electromagnetic forces known as quantum electrodynamics, and the theory of the 'weak' force that creates some forms of radioactivity and powers the sun's nuclear fusion, looked something like two sides of the same coin (see *Which is Nature's Strongest Force?*). Weinberg and Salaam showed that this was indeed the case, and unified them into the 'electroweak' theory. There was a problem, though. That theory required that a couple of as-yet-unseen particles, dubbed the W and Z bosons (a boson is a particle that creates a force), be added to the so-called particle zoo.

Rather embarrassingly, these two particles had mass. That seemed wrong, because the most famous boson is the photon, which creates the electromagnetic force, and the photon has no mass. If the photon and the W and Z bosons do the same kind of job in a unified theory, there ought to be a kind of 'symmetry' between them. The fact that there isn't, because of the masses of the W and Z bosons, leads physicists to suspect that something is breaking that symmetry, in the same way that adding a weight to a carefully balanced kitchen scales will upset its delicate balance. But what was that weight? This is the question to which Peter Higgs offered his field as an answer.

By 1967, Weinberg and Salaam had incorporated the Higgs field into their electroweak theory. In 1983, at CERN, the W and Z bosons were seen, exactly as Weinberg and Salaam had predicted. It was a triumph, the closing refrain of the particle physics odyssey. Except for one tiny detail. No one knew whether the Higgs field was really there.

Hunting the Higgs

You can envisage the Higgs field in various ways, but one is to rub your finger along the groove in a sheet of corrugated metal. It feels smooth and your finger runs without resistance. Now pull your finger across the grooves. It is much rougher. In the standard model of physics, this is how things are for the W and Z bosons. While the photon always moves along the Higgs field's grooves, the other two move across them, encountering a resistance that translates into mass.

It's an elegant idea, but it needs proof. And the only way to find that proof that the Higgs field really does provide a directional 'grain' to the universe, felt by W and Z bosons but

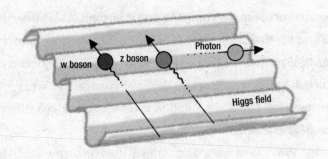

The effect of the Higgs field

not by the photon, is to find the particle that the field produces. Every field has its own particle. The electromagnetic field has the photon, the gravitational field has the graviton (though no one has ever seen one), and the strong interaction comes through the gluon. The Higgs field, according to received wisdom, endows things with mass because of the Higgs boson. The question is, is received wisdom to be trusted?

Physicists do not have unlimited confidence in their theory of particle physics. In some ways it is hugely successful. We have predicted the existence of particles that have always been found – and in many cases the theory even told us exactly where to look. Physicists measure particle energy in electronvolts or eV; an electron would gain 9eV of kinetic energy when pulled by the voltage across the terminals of a 9 volt battery, for example. Salaam and Weinberg told CERN researchers that if they smashed particles together at 80 and 90 gigaelectronvolts, or GeV, they would find the w and z bosons. And that's exactly what happened.

However, the standard model does not predict everything. The fact that 26 of its fundamental constants have to be found in experiments then written into the equations, for instance, is a little frustrating. Some particles had to be found by trial and error too. There was a 20-year gap between the prediction that the fundamental particle the 'top quark' must exist and the moment we finally found it. That's partly because the theory gave us no idea where to look (it turned up at 170 GeV). Unfortunately, we were in the same boat with the Higgs boson. It should be there, but no one knows where 'there' is. And so we kept building bigger and bigger atom-smashers in the hope that we'd eventually get to the right energy.

Smash and grab

All this is not as desperate or random as it might appear. Smashing atoms together has a solid history as an experimental tool. That is, after all, how Ernest Rutherford discovered the atomic nucleus. In 1909, he decided to test the 'plum pudding' model of the atom, which suggested its positive and negative charges were mixed together. Rutherford fired a beam of alpha radiation – essentially the nucleus of a helium atom – at a thin sheet of gold foil. Most of the alpha particles were unaffected, but some were deflected wildly. From his results, Rutherford deduced the existence of a tiny region of concentrated positive charge at the centre of the atom which caused the occasional wild deflections. Nuclear physics was born.

Since Rutherford, we have built up a roster of ever-larger particle accelerators in order to probe the complexities of the nucleus, culminating in the current state of the art: the Large Hadron Collider at CERN. Though it might seem like it from the media coverage, the LHC is not the first particle accelerator to be raised up as a detector for the Higgs boson.

Because we had no way of telling at what energies the Higgs boson might be found – though the standard model suggests 96 GeV as a likely target – we have been hoping to stumble across it for many years. But particle accelerator after particle accelerator has been hailed as the bright new hope and we are still not there.

The first one with a serious chance was the Large Electron Positron (LEP) collider at CERN (see *Why is There Something Rather Than Nothing?*). Housed in a circular tunnel 27 kilometres (16.5 miles) in circumference, LEP accelerated electrons and positrons to close to the speed of light. A ring of 4,600 magnets guided these particles in a circle that extends out across the Swiss border into France, to the foothills of the Jura Mountains, with the electrons going in one direction and the positrons in the other. The magnets could be tweaked to guide these beams into each other, releasing a cascade of particles from any collisions. Four vast detectors, each the size of a small house, would pick up the tracks of these particles. The experiments would run for hours, with potential collisions every 22 millionths of a second. Then the scientists had to examine the output of the detectors, and try to work out what had happened as electron and positron smashed into one another.

A glimpse of the Higgs

LEP, which came into operation in 1989, accelerated particles to 45 GeV, enough to produce the Z boson. Later upgrades enabled it to produce the W boson too. By the time it was scheduled for shutdown, LEP was operating at 209 GeV. But just before that, in September 2000, it produced a tantalizing glimpse of something that looked like the Higgs boson.

The observation was made at collisions involving energies of a whisker under 115 GeV, which made sense in the standard model's view. Unfortunately, there were simply not enough of the observations to make it a statistically significant result. The only conclusion was that, using Einstein's $E = mc^2$ energy mass equivalence, the Higgs boson was heavier than 114 GeV.

The mass of the Higgs boson is tightly linked to the mass

High energy particle accelerators

NAME	PARTICLES COLLIDED	ENERGY
Stanford Linear Collider (SLC)	electron, positron	100 GeV
Large Electron Positron Collider (CERN-LEP)	electron, positron	200 GeV
Relativistic Heavy Ion Collider (BNL-RHIC)	heavy ions	200 GeV
Tevatron (FNAL)	proton, antiproton	2 TeV
Large Hadron Collider (CERN-LHC) in Geneva	proton-proton; ion-ion	14 TeV

of the top quark and the mass of the W boson. As scientists pinned down these masses ever more precisely, the range of energy scales over which the Higgs boson could appear became narrower. A constraint on the mass of the W boson brought the most likely Higgs in at 153 GeV, and the race was now on to find it. In 2009, scientists at Fermilab announced they had a 50:50 chance of spotting the Higgs boson before the end of 2010. They didn't manage it, leaving LHC researchers to get there first. This new collider, the world's most powerful machine, occupies the tunnel vacated by the LEP in 2000. It will accelerate protons and antiprotons to a stunning 99.9999991 per cent of the speed of light. The particles will eventually crash together at 14 TeV (teraelectronvolts). With all that energy concentrated into beams only a thousandth of a millimetre across, it's a prospect that had some people worried that the collider could produce unexpected and catastrophic effects.

The issue is the high concentration of energy in the LHC's colliding particles. The energy is not enormous – it is something around the kinetic energy of a small insect. But it is concentrated into a very small region of space. We know from Einstein's theory of relativity that energy warps space. In some models of the universe, where there are more dimensions of space than the three we experience, such a concentration of energy can produce tiny black holes, where the extreme bending of space creates what is effectively a rip in space and time.

In this scenario, the black holes generally disappear in a fraction of a second and pose no threat. There is a tiny, but finite possibility, though, that they can grow to a significant size and become a real danger. Or that's the scaremongers' theory. The reality is far more prosaic. The scare stories have prompted researchers to go through the theory with a fine toothcomb. They unanimously conclude that the probability of disaster is infinitesimally small.

Perhaps more important, though, is that whatever the theory says is possible (but highly improbable), we actually have some experimental results that also bear on the discussion. In our upper atmosphere, charged particles from outer space are causing higher energy collisions than those that will take place at the LHC. They are happening at the rate of 10,000 billion LHC collisions per second. As the LHC's safety report puts it, 'over the history of the Universe, Nature has carried out the equivalent of 10^{31} LHC projects'. That's 10,000 billion billion billion LHCs, with no sign of a black hole opening up and eating the Earth. On that basis alone, there is no reason to think that the LHC poses any danger to the future of humanity.

Technical issues caused massive delays to its start-up, but the LHC remained the best hope for detecting the Higgs particle. It took years before the LHC produced any meaningful data, though. Its detectors are enormously complex, and needed an unprecedented amount of calibration. Once that was done, the experiments were started. So what if the

Higgs hadn't appeared in the LHC? It may sound shocking when you consider the £2.6 billion cost of the collider, but physicists were sanguine about that eventuality.

A sign of supersymmetry

The accepted view is that, if there is no Higgs boson, the standard model of physics will crumble and fall. To explain the asymmetry between the massless photon and the massive W and Z bosons requires a Higgs boson, or something like it. But even here, there is wiggle room. Some physicists claimed, for instance, that we have oversimplified our understanding of what the Higgs signature might look like. If things are more complicated, because of a theory that goes beyond the descriptions offered by the standard model. This theory is called 'supersymmetry'.

In supersymmetry, every particle has a 'superpartner' that is a heavier version of itself. The electron is partnered by a selectron. The quarks have squarks. And so on. This creates a much more complex particle zoo than we might perhaps like, but the idea has teeth. Most importantly it solves numerous difficulties with the idea of 'unifying' all the forces of nature. The link between the electromagnetic and the weak forces, for instance, hints that all the forces evolved from one superforce just after the Big Bang. As things cooled down to lower energies, the superforce split into the forces we now recognize. Supersymmetry suggests that there are perhaps as many as five particles to be associated with the Higgs boson. So what did this do to the search? Well, predictably, it made it more complicated than we would like.

The Higgs trail

Each of the supersymmetric Higgs particles would produce a different decay signature in the LHC's detectors. Each signature takes the form of a trail of particles that exist at certain energies then decay into other particles with particular properties. Though that might sound easier to spot than just one Higgs, the truth is that this complexity makes it easier to miss – or to get a false positive from the decay of other, non-Higgs particles.

The energies of these Higgs particles are such that there was also a chance they might be spotted in Fermilab's Tevatron, the most powerful accelerator in the United States. They might even have been there in the data coming out of the LEP collider; maybe we just weren't looking in the right place. It's still an open question. But spotting any signature of supersymmetry will be a boon. In fact, CERN researchers expected the LHC to find evidence for supersymmetry – the first evidence for supersymmetry – before it found the Higgs boson.

In the end, we believe we have now found the Higgs boson at 126 GeV, but the root of mass is still not fully exposed; we still don't understand why particles have the mass they do –

why, for instance, the top quark has a million times the rest mass of an electron. The Higgs does reconcile the existence of mass with the way the weak force works, but why it gives so much mass to the quarks remains a mystery. What's more, the mass of the quarks in a proton, added to the energy that holds the quarks together, still doesn't add up to the mass of the proton.

And there is yet more bad news for mass-hunters: we have no explanation at all for the electron's mass. Whatever does eventually happen in our accelerators, it seems clear that the God particle is much less important to the future of physics than its name might lead one to believe. Seeing it – or something like it – is thrilling, but we may well discover that our God particle has feet of clay.

Am I unique?

The limits of our universe and the search for parallel worlds

If you have ever wondered what makes you you, or whether there is some unique, pre-determined purpose and path for your life, you have asked one of the biggest questions that physics could conceivably answer.

It is a question that writers play with all the time. Tales of other worlds, reachable from ours, abound in literature. The idea is a science fiction staple, but it also forms a central theme in books for children. There is the fictional world of Narnia in the C.S. Lewis series, for example, and the classic Lewis Carroll story of *Alice in Wonderland*.

But these books tend to assume uniqueness for their heroes and heroines, who reach a parallel world without losing time in their own world. This is no doubt to do with the limits on our consciousness, which tells us there is only one 'me': I can be in only one place at a time. But if we set aside the Zen-like problem of consciousness, and what 'I' is, the answer to our question is almost certainly a straight-forward no: you are not unique. How we get to that answer, however, is far from straightforward.

There are three reasons why you might not be unique, and all of them are central to our view of the universe. One is to do with the physical extent of the universe, and whether it has an edge. The second has to do with something that Einstein called the 'biggest blunder' of his life, and reaches out from the first moments of creation to raise questions about our infinite future. The third probes the essential nature of the quantum world. If 'am I unique?' seemed like a silly question at first, it doesn't now. The issue of whether there really is another you somewhere out there is actually the same as asking how much we know about the universe.

The simplest route to an answer is through an examination of the size of the universe. Here physicists have three possibilities to choose from. Perhaps the universe is infinite in extent. Or it could be finite but, like an ant on a tennis ball, we can never reach the edge. The third option is that the universe is finite, and its geometry is such that you could fall off an edge.

If the universe is infinite, then there is good reason to think that you are not unique. Though it would contain infinite numbers of worlds, and thus infinite numbers of worlds with Earth-like life, it seems that there are only so many ways a set of molecules can be configured to give a living being. That would mean that, somewhere, there is a carbon – pun intended – copy of you.

Of course, you will immediately counter this suggestion, saying that, even if all the molecular structures are identical, it wouldn't make it *you*. There is the issue of memories and experience, but, apart from that, what is *you*, exactly? We are getting ahead of ourselves – at this stage we don't even know if it's a question that has to be faced. The question at hand is now, is the universe finite or infinite?

Infinity and beyond

Scientists and philosophers have long pondered the size of the universe, but for most of history, it has been considered finite. Around AD140, Ptolemy conceived of the cosmos as a finite sphere centred on the Earth. Only in 1576 did anyone suggest otherwise. That was when the English astronomer Thomas Digges put forward the idea of an infinite universe populated by stars similar to our sun. Digges was more fortunate than the Italian philosopher Giordano Bruno. When Bruno suggested something similar a few years later, he ignited the fury of the authorities of the Roman Catholic Church, who sent him to be burned at the stake.

We are still none the wiser as to the extent of the universe. Observations of the cosmic microwave background radiation, the echo of the Big Bang, seem to indicate that the universe might be finite. The most popular explanations of anomalies in the spectrum of this radiation suggest a limit to the size of the cosmos, but there are plenty of competing explanations. So, as we are unsure about whether the universe is infinite or not, we cannot say whether there is another you on a faraway world. Perhaps our second possibility, the one involving Einstein's biggest blunder, can shed more light on the issue.

A froth of universes

This possibility begins with something known as 'eternal inflation theory', which involves a succession of universes bubbling out then pinching off from one another. Though it sounds odd, there is quite some evidence for this as a natural, ongoing scenario. The idea was born with the discovery of an anomaly that haunted physics in the 1970s. A decade after the 1963 discovery of the cosmic microwave background radiation, few people doubted that the universe had begun with a 'Big Bang'. The term had been coined by Fred Hoyle, one of the idea's most strident critics, as a way of deriding the idea that the universe exploded into existence, but the evidence was good, the name catchy, and – probably most important – it fitted nicely with the dominant religious views of creation. There was one problem, though. The universe we saw couldn't be explained by a big bang alone.

For a start, relativity tells us that space and time curve when in the presence of energy and matter (see *Why Does*

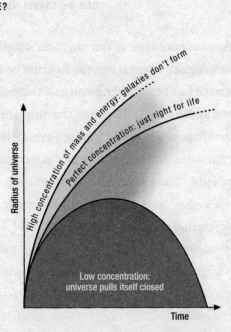

The delicate balance that gave rise to the universe

an Apple Fall?). That will have a profound effect on our universe, altering its overall geometry. The consequences of this geometry change depend on how much matter and energy there is. In high concentrations, space and time would curve catastrophically, closing up the universe. In low concentrations, the expanding power of the Big Bang would have dominated the shape of the early universe, throwing everything in it so far apart that stars and galaxies (and habitable planets) would never form. However, our universe was perfectly set up with a 'flat' geometry that allowed us to eventually exist. The question is, why should it be so perfect?

This 'flatness problem' is not the only tough question facing cosmologists. There is also the 'horizon problem'. This stems from the fact that the temperature at opposite ends of the universe appears to be the same. The only way for that to happen is if heat has been distributed evenly through the universe, but we know that the universe is too big for that to have happened. Heat is carried by photons, which are particles of radiation. Even though photons travel at the speed of light, there has not been enough time for photons to move throughout the universe, carrying heat from one extreme to the other, so that the cosmos no longer has hot spots.

Inflation to the rescue

At the beginning of the 1980s, physicists solved these two problems with a single stroke. The solution was called 'inflation', and it suggested that, just after the Big Bang, the universe went through a period of super-fast expansion. Although no one knows how or why it might have happened, a period of inflation is still the best answer to the problems cosmologists have with the Big Bang, explaining the spread of heat and the flatness of the universe. It also provides a path to a second you.

People have been playing around with possible mechanisms for inflation for nearly three decades now. The most popular ones suggest that inflation is a never-ending story. If a tiny point of space–time blew up once, it can do it again. According to these chaotic inflation theories, the fluctuating energy inherent in empty space can inflate a whole new universe from anywhere within our own space and time. In a process reminiscent of something in Willy Wonka's chocolate factory, new universes are bubbling up from old ones all the time. The mouth of each one eventually pinches off, separating it from its parent for ever.

Though this does seem fantastical, the scenario got a big shot in the arm when string theorists seized on it as an idea that would solve their own set of problems. String theory is an attempt to create a 'final' theory of physics that unites Einstein's relativity with the strangeness of the quantum world. The basic idea is that all matter is composed of tiny vibrating loops of energy; the frequency of the vibrations determines what kind of matter shows up. When string theorists tried to calculate the kind of universe that this would create, they were hoping they'd end up with one that looked and behaved rather similar to ours.

They didn't. However hard they tried, they couldn't create a single string universe that matched the one we live in. Instead, they created thousands, each one endowed with a different set of characteristics. The problem was compounded by the 1998 discovery that the expansion of the universe was speeding up. Although we expect the universe to be expanding still – the Big Bang's influence is still strong – it should be slowing down as the gravitational pull of everything in the universe works against the expansion. If the expansion is speeding up, some unknown force is at work.

It didn't take long for physicists to work out that the energy associated with this acceleration makes up approximately 70 per cent of the total mass and energy in the cosmos. Call it what you like – physicists call it dark energy – but that's an awful lot of stuff to not know about.

Einstein's mistake

The best answer to the dark energy mystery lay with a mathematical term that Einstein had crowbarred into his original equations describing the universe. Einstein didn't know anything about a Big Bang, and thought the universe should be static, not expanding. Unfortunately, his equations created an unbalanced universe, so he inserted this term, known as the 'cosmological constant', to create a neat, static universe. After the discovery of the Big Bang, he called it his 'biggest blunder'.

With the discovery of dark energy, however, the cosmological constant came right back into fashion. This term, it was thought, could explain why the universe was expanding ever faster. But it didn't – in spectacular style. The calculated value for the constant came in at around 10^{120} times the measured value. That's 1 with 120 zeroes behind it: even physicists have labelled it the most embarrassing mismatch between theory and experiment in the history of science.

But string theory has an answer. Don't expect to understand why a universe is as it is; just glory in a multiplicity of diverse worlds. Chaotic inflation says they all exist, and so does string theory. Yes, we live in a universe with an inexplicably small cosmological constant, but why do we think we should be able to calculate the values of the constants of nature from scratch? They simply are what they are – and they are different in every one of the vast landscape of universes that string theory predicts to exist.

The current thinking at the frontiers of theoretical physics is that, rather than being a problem, the inexplicable value of the cosmological constant is proof that string theory is on the right track. It might seem like twisted logic, but if the string theorists are right, it does provide the route to another you. The vast landscape of universes bubbling out from each other via chaotic eternal inflation has no end. Though their constants of nature are, effectively, random, some of them will be identical to ours. That means planets will form, stars will appear and cluster into galaxies, and elements such as carbon will be synthesized in the burning cores of those stars. Life will emerge and, in some cases, so will humans.

And there 'you' are. The chance that something with exactly your genetic make-up will appear on a blue-green planet somewhere in another universe seems infinitesimally small. But that tiny probability is converted into certainty when we allow the existence of an infinite number of universes. Not that you two could ever meet. When a new universe bubbles out and pinches off, contact is lost for ever. You are trapped in your own space and time; your twin is in a separate and unreachable sphere.

Worlds without end

The weirdness of a bubbling universe pales beside your third and final chance for multiplicity, however. Quantum theory provides not just a chance of there being another you, but an argument that there are a near-infinite number of you. The twist is, each one has made a different choice in life. This is the 'many worlds' interpretation (MWI) of quantum theory, and it is truly mind-bending.

There are a number of interpretations of quantum theory, and each one has to explain the inexplicable. The theory allows for quantum particles – atoms, electrons, the bullets of light energy known as photons – to exist in more than one state at any one time. This phenomenon is known as super-position, and it is a profound mystery. An electron can spin clockwise and anticlockwise at the same time, for example. A photon can be simultaneously here and there. An atom can hold two different energies.

In the classic demonstration of superposition, physicists fire electrons at a screen that is scored by two narrow vertical slits. The stream of electrons is so slow that there is only one particle in the apparatus at any one time. To our thinking, the electron will go through one or other of the slits. Place a phosphorescent screen, rather like the screen of a cathode ray tube television, behind the slits, and we should see two sets of glowing dots where the electrons land: one behind the left slit, and one behind the right slit. We don't. We see a series of glowing bands known as an interference pattern.

Interference is something we associate with waves. Ocean waves interfere with each other: when the crests meet, they reinforce and the water piles up higher. When two troughs – effectively, negative quantities of water – meet, an even deeper trough is the result. When crest meets trough, they cancel out to give flat water.

The same is true for light, as Thomas Young demonstrated two centuries ago. Young was demonstrating that light is a wave, overthrowing Newton's particle theory of light. In an arrangement like the double slit experiment described above, but with light passing through the slits, Young's screen showed a series of light and dark bands, something that could only be achieved if the slits both acted as secondary sources of light, with the two emerging light waves interfering.

Going back to the single electron in the double slit experiment, then, how to explain an interference pattern? How can there be interference when there is only one particle? The answer is that, although we think the electron has to go through one or other of the slits, it actually goes through both. An electron might be a particle, but it is also a wave.

There is no easy resolution to this paradox, and the world's greatest minds have debated it endlessly since quantum theory was invented. In the 1950s, however, Hugh Everett came up with a radical new take on the problem. At the time it was much derided, but today it is gaining support. The idea is simple. Every time a quantum particle faces a choice, new worlds are created – worlds in which every option is realized.

A new universe at every moment

It's easy to see why scorn was poured on Everett's idea: who can stomach the notion that a world is created every time a photon is spat out by a star, or absorbed by an atom in a human retina? These are both quantum events, where one quantum particle is absorbed by another. Can we really believe that just looking at the heavens forces a new universe into existence? Everett left physics shortly after publishing this idea, but it has nevertheless found a series of champions. That is largely because, strange as it seems, it actually offers a reasonable solution to the strangeness of the quantum world.

In Everett's many worlds interpretation, the electron doesn't form a superposition state when faced with a choice of two slits, but splits the world into two. In one world, it goes through the left slit. In the other world, it goes through the right hand one. Though we have no consciousness of the different worlds, quantum particles such as electrons feel their influence from across the divide. The pattern we see results from interference between electrons in different worlds. In this view, what we think of as reality is just one of an infinite number of realities, each one slightly different from the next. And each one will contain a version of you.

The MWI seems to have a slow-growing following amongst physicists; a 1995 poll of physicists attending a conference on quantum theory found that 60 per cent believed it to be the correct interpretation of the theory. Such polls are unscientific, though, and not an indication of the 'rightness' of everything. Which is why, if you are really intent on finding out the truth about that other you, you have to consider a radical proposal: quantum suicide.

Don't try this at home, but the protocol of this experiment is fairly simple; it could even be done using currently available technology. You hold a loaded gun to your head, but rig it up so that pulling the trigger prompts a measurement on a quantum particle – determining the spin of an electron, for example. If the result is 'clockwise', those standing around watching hear a click. If it's 'anticlockwise', they see the gun fire. Not a pretty sight.

But here's where perspective becomes everything. If Everett was right about the existence of many worlds, there will always be a world in which the gun doesn't fire. Your conscious existence will, therefore, never know of the gun firing. After a dozen clicks you'll be convinced that quantum suicide is actually a route to appreciating not only the multiplicity of your existence, but also your immortality. Not that you'll be able to share that viewpoint with anyone. What's more, you can have your cake and eat it. You have found that other you, but you can also leave it behind, hopping from world to world like Alice in a quantum wonderland.

Can we travel through time?
Where relativity meets science fiction

'Scientific people know very well that time is only a kind of space. We can move forward and backward in time just as we can move forward and backward in space.' This might sound like a claim from the future, or at least the present, but it comes from the past.

It is spoken by the Time Traveller in H.G. Wells's *The Time Machine*, which was published in 1898. The truly remarkable thing is Wells's prescience: nearly twenty years passed before Albert Einstein published the theory that made such time travel theoretically possible – and even then it took years before anyone noticed.

Oddly, Wells's Time Traveller only ever travels to the future. Now, however, we know that the laws of physics allow for travel forwards and backwards in time. If you can handle ideas such as infinitely long rotating cylinders the size of galaxies, wormholes held open with exotic forms of negative energy, and having to choose between never having been born or losing your free will, then you might just be able to handle the science of time travel. As thrill-rides go, it's a little bumpy. But, given the prize, it's definitely worth it.

Loops in time

Time travel is so fascinating because we are trapped by time. We cannot choose how we move through it, as we do with the other dimensions. But Wells's idea that if we only knew how, we might be able to treat time just like space, was right on the money.

In 1915, Einstein published his general theory of relativity. This described the universe as a four-dimensional fabric made up of three dimensions of space, and one of time. Every piece of matter and energy in the universe warps the fabric, changing the shape of the universe in a way that causes matter and energy to experience the pull we call gravity. The sun, for instance, creates a kind of well in the fabric, into which nearby planets would fall, if it were not for their momentum. The result is that the planets orbit the sun in the same way that, in a casino, a speeding ball orbits the centre of a spinning roulette wheel.

It is easy to imagine how the undulating landscape of gravity affects motion through space. But the same is true of motion through time: this undulates too. Pack enough mass and energy into a small enough region of space and you can even bend time into a loop – it is rather like rolling up a sheet of rubber so that the ends meet and you can walk around the surface without ever reaching an endpoint. In this configuration of the universe, a moment repeats itself endlessly.

The first person to notice that general relativity allows the creation of loops in time was the Austrian mathematician Kurt Gödel. In 1949, in a review article describing how the invention of relativity had changed our perception of the universe, he wrote that it 'is possible in these worlds to travel into any region of the past, present and future, and back again, exactly as it is possible in other worlds to travel to distant parts of space.'

Gödel had solved Einstein's equations and found that, if the universe is rotating, time can flow in loops. He was alarmed by this, and being a close friend and colleague of Einstein, Gödel showed him the result. Einstein said he too was 'disturbed' by the possibility. 'It will be interesting to weigh whether these are not to be excluded on physical grounds,' he wrote in a reply to Gödel's paper. Gödel seemed to have a similar view: something, he suggested, must stop such things happening. The universe surely cannot be allowed to have people travelling through time.

Return to the past

In some ways, Einstein need not have worried. Gödel's work was solid, but useless. The motions of the galaxies tell us that our universe is not rotating, so loops in time will not naturally exist. If we are to build a useful time machine, we will have to create those loops for ourselves.

But we do have ideas for how to accomplish that. The first came in 1976, when Frank Tipler of Tulane University in New Orleans, Louisiana, drew up the blueprints for a time machine. Tipler showed that an extremely massive and infinitely long, fast-rotating cylinder would warp the fabric of the universe enough to create a loop in time.

Again, though, this has little future as a time machine. It is certainly not the kind of thing that Wells envisaged: his Time Traveller builds a time machine that fits into his house. Infinitely long cylinders are hardly likely to fit in any factory, however large. There is another option: use time machines that nature has already built. In 1991, Princeton astrophysicist J. Richard Gott showed that the universe might contain material that could act as the raw material for a time machine. The material is a super-dense strand of 'cosmic string'.

According to some theories of how the universe formed, cosmic strings would have formed in the earliest moments of creation, and might still be hanging around the universe today. They are, essentially, defects in space, something like scar tissue that formed when the universe was going through a period of rapid change. A cosmic string is a fearsome beast: though less than the width of an atomic nucleus in diameter, it stretches across the universe. Unsurprisingly, turning one into a time machine is not for the faint-hearted. For a start, you need a pair of them.

The extreme density of each of the strings warps space–time in a way that means you can create a loop in time by placing them side by side, then moving them rapidly apart. Travel in a loop around these moving cosmic strings and every time you return to where you started, you will find yourself at an event in your past. Gott compares it to an Escher drawing. Just as Escher warped perspective to create geometrically impossible effects, the strings warp the geometry of the space–time around them so much that it no longer follows the rules that are familiar to us.

The same effect, he has pointed out, might be achieved by firing super-energetic particles towards each other so that they missed by only a tiny distance. Their energy would warp the space–time around each particle, and when those warped space–times met, they could form a loop in time. It's not a loop that you could enter and walk round, however. Much more interesting – and feasible – is the wormhole time machine drawn up by American astrophysicist Kip Thorne.

Into the wormhole

You will have almost certainly heard of wormholes because they are a staple of science fiction. But that is entirely justifiable: although they have occupied countless hours of research time, this method of time travel was actually inspired by a science-fiction story. When the cosmologist Carl Sagan was writing his novel *Contact*, he wanted to find

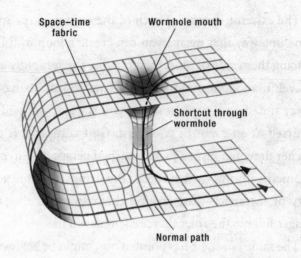

The shortcut path through a wormhole

a plausible way to send his heroine to Vega, a star 26 light years away, in an instant. Sagan asked Thorne's opinion, and Thorne set to finding a solution. He found it in a paper that Einstein had written with Nathan Rosen in 1935.

They had solved a problem associated with black holes, the remnants of stars that have collapsed under their own gravity. At the core of a black hole is a 'singularity', a breakdown in space and time. Einstein and Rosen had imagined this central core connecting with another region of space–time. This is the Einstein–Rosen bridge. Thorne soon realized this could be the answer Sagan needed.

Think of a railway engineer trying to lay track to the other side of a hill. You could lay it up one side and down the other. You could divert the track around the hill. But if there was a tunnel through the hill, that would be a shorter, more direct route. Because time and space are so closely related in relativity (physicists put them together and talk about the fabric of the universe as 'space–time'), you can do for time what railway engineers already do for space: you can engineer a short cut. Subsequent analysis of the geometry of these shortcuts through space–time has shown that they would work for time travel.

Short cut through time

To specify a point in space–time, you give a position and a time: St Paul's Cathedral at noon today, for instance. If this is one mouth of the wormhole, the other might be St Paul's Cathedral at noon yesterday. Walk into the mouth today, and you would walk out at the same point in space, but 24 hours earlier. It might take you some finite time to move from the wormhole's entrance to its exit, but that needn't be a problem. In theory you could jump in at one mouth, emerge in the past and hang around to watch yourself jumping into the wormhole.

It's not quite as easy as that, of course: there are several obstacles to overcome. One is, where do you find a wormhole? Though they exist as solutions to Einstein's equations, there is no evidence they exist naturally. There is a remote possibility that we could create one by colliding fast-moving subatomic particles. Various theoretical ideas indicate that their highly concentrated energy might warp the fabric of space–time enough to tear a hole in it. But even then we would not be in control.

Space–time is like elastic: it doesn't like to be stretched. The way space–time tears to create a wormhole creates an energy imbalance that tends to pull the mouth of the wormhole shut. The only way to keep a wormhole mouth open, physicists reckon, is to pack it with 'negative energy' that pushes against the natural closure. Though it is possible that a material carrying negative energy exists, we have no idea what it might be or where we might look to get some. Assuming we can keep the wormhole open, who is to say that the hole would bridge to another area of space–time? And if it did, would it be where we wanted to go?

The best solution to this problem (given the existence of a wormhole and fantastical technological abilities with negative energy) seems to involve anchoring one end of a wormhole to a neutron star. A neutron star is an amazingly dense object. Though only around 20 kilometres (12.5 miles) across, a neutron star weighs more than the sun. In Earth's gravitational field, one teaspoonful of neutron star material would weigh a billion tons.

This concentration of mass has a profound effect upon the space–time around a neutron star: it warps it severely. One of the results is that time slows down in the vicinity of a neutron star. Near a neutron star, time runs at about

30 per cent of the speed it runs on Earth. Tow one end of a wormhole to a neutron star, then, and let the other end sit in empty space, and a time-shift would develop between the two mouths of the wormhole. In theory, that means you could enter the wormhole at a time after you emerged at the other end.

Protecting the flow of time

OK, so none of this is easy. Why not? It's not because creating a time machine breaks some fundamental law of physics. A better suggestion is that we operate by rules that 'conspire' against time travel. Perhaps, as Gödel and Einstein suggested, the disturbing implications of someone travelling into their past are a way of alerting us to the fact that something in the universe makes it impossible.

As every Hollywood screenwriter knows, time travel to the past certainly throws out some weird and wonderful dilemmas. The classic one is known as the 'grandfather paradox'. What if you went back in time, and killed your grandfather when he was a young boy? That would mean one of your parents was never born – would it also negate your own existence? Would you be rubbed out of reality?

There are three possible solutions to this. The first, and most plausible to physicists who think about time travel a lot, is known as the 'chronology protection conjecture'. It was conceived by Stephen Hawking in 1992, and suggests that some as yet unknown aspect of the natural world will kick in if the flow of cause and effect is ever threatened. Basically, the laws of physics conspire to protect the past. It's a neat idea.

Everywhere physicists look, there certainly seem to be unanticipated factors weighing down any attempt to create a time machine. There is the need for negative energy for wormholes. Gott's cosmic string time machine seems to suffer from the setback that the universe conspires against you ever assembling enough mass in a small enough location. There are even hints that quantum versions of time machines, proposed to incorporate physics that is not yet properly understood but will one day have to be taken into account in considerations of time travel, have their own brick walls.

And yet Hawking's chronology protection conjecture is still just an idea – a way of sidestepping awkward questions about the grandfather paradox without forcing physicists to give up looking at time travel. The second possibility for protecting your grandfather comes from the quantum world, where weird problems can always find similarly weird solutions. In this case, the idea is quite simple: everything that happens creates a new universe that has no connection to any other universe.

This idea, dreamed up by Hugh Everett in the 1950s, is known as the 'many worlds hypothesis' (see *Am I Unique?*) and is used to resolve a long-standing problem in quantum theory. Its application to the paradoxes of time travel is equally simple – and equally exasperating. If you go back in time and kill the young boy you think to be your grandfather, you enter a different, parallel world, one where your only existence is that of the time traveller, a totally separate existence from the grandson. There is no 'other you' whose existence can be called into question. Paradox resolved.

But, again, not in a particularly satisfactory way. The third idea is simply that we do not have the control over the external world that we think we do. This approach to the paradox says you do not have free will and would not be able to kill your grandfather even if you wanted to. This is a complicated area, and raises philosophical questions that physicists are not equipped to answer. If they really want to know how the grandfather paradox plays out, they need to get on and build a time machine.

To the future

All of this seems like it is leading to the conclusion that we cannot travel through time. But nothing could be further from the truth. We know time travel is possible because we have already done it.

The *Apollo* astronauts who flew on rockets to the moon and back became, effectively, the world's first time travellers. The world's greatest time traveller is Russian cosmonaut Sergei Krikalev, who circled the Earth for about 800 days at

17,000 miles an hour. Krikalev is now one forty-eighth of a second into the future.

You don't even have to be a cosmonaut to travel through time. Experiments with highly sensitive atomic clocks flown around the Earth have shown that they moved into the future. A round-the-world trip on an aeroplane might gain you something on the order of a few billionths of a second. Why? The answer lies with Einstein's first relativity theory: special relativity.

Special relativity (see *What is Time?*), which was published in 1905, says that the passage of time for any person or object is relative, and depends on motion. If you blast off on a rocket bound for Alpha Centauri, say, your watch will run slow compared to the clocks back on Earth. If your rocket travels at close to the speed of light, that difference in measured time could be profound. In a long but fast return journey, it is possible that you could arrive back on Earth just a few years older, but find that everyone who stayed home has aged much more.

In this scenario, if you have a twin, they would no longer be the same age as you. This bizarre result, known as the twin paradox, is fully allowed by the laws of physics. The truly remarkable thing is that this difference in the passage of time means the travelling twin has travelled into Earth's future. When you return from your travels in space, you find that you have also travelled in time: more time has passed on Earth than has passed for you. We can conclude therefore that we can indeed travel through time – and some humans have already done so. However, this travel into the future is relatively easy. It is travel to the past that is proving so difficult. Will we conquer these difficulties? Only time will tell.

Is the Earth's magnetic shield failing?

Drifting poles, the planet's churning core and the threat to life on Earth

Can we avoid the fate that befell Mars? The Red Planet's magnetic shield failed, and its atmosphere was blown away by the sun, leaving it a barren, sterile wasteland. Is Earth heading the same way?

Earth's magnetic field, known to scientists as the magnetosphere, has been an integral part of the biosphere since life on the planet began. Bacteria, plants and animals are known to be affected by its orientation. Many species of birds would quite literally be lost without it – it is the cornerstone of migration strategies that allow them to escape harsh northern winters, for example.

Humans cannot consciously sense magnetism in the same way as many animals, but we still gain enormous benefit from the Earth's field. Not only does it appear to keep our atmosphere in place, it also protects us from intense solar radiation and electrical storms that would otherwise play havoc with our electrical grids, satellites and aircraft communications. If Earth's magnetic shield is failing, we need to know sooner rather than later.

Drifting poles

We may never know which human civilization was the first to make explicit use of Earth's magnetic field. Until relatively recently, it was thought to be the Chinese, who used magnetic minerals known as 'south-pointing fish' to align their buildings in accordance with the principles of feng shui. However, reliable evidence for this practice dates back no further than 400 BC which means the oldest magnetic artefact is most probably a piece of the mineral magnetite found in Veracruz, Mexico, home of the Olmec.

The Olmec were, it is thought, the first civilization of the New World, existing between 1000 and 1400 BC. The piece of magnetite, unearthed in the early 1970s, had been fashioned into a bar that would offer little friction when set on the ground, and scored with a groove in the middle of one end. It looks, to all intents and purposes, like a compass needle.

When physicist John Carlson reported the discovery of the Olmec magnetite, he pointed out that the Olmec people constructed their buildings on alignments 8 degrees west of north. This, he said, was 'a curiosity'. But taken with other evidence gathered over the subsequent centuries, it is more than a curiosity – it is evidence that the Earth's magnetic field is far from constant. And that is why we think it may currently be failing.

A compass that doesn't point north

Modern measurements of the Earth's magnetic field only began two centuries ago, but we do have older evidence of shifting fields. Examine the orientation of over a hundred Danish churches built during the 12th century, for example, and you'll see that they sit around 10 degrees off from today's magnetic east–west line. As with the Olmec buildings, it is highly likely that when these churches were built, compasses pointed in a different direction than today.

A more reliable account of Earth's magnetic field began at the start of the 19th century, when Alexander von Humboldt took field measurements while travelling in the South Atlantic. He found that the intensity of the field decreased in this region. In 1804, von Humboldt reported his finding to the Paris Institute, but counter-claims soon emerged, throwing the issue into confusion. Von Humboldt eventually took the matter to the German mathematician Carl Friedrich Gauss, and asked for help in constructing an atlas of magnetic observations. Gauss, a polymath who had made important discoveries in disparate scientific fields, was already investigating terrestrial magnetism, and was only too keen to help. By 1840 he had written three significant papers on magnetism – including a way to define the Earth's field –

and constructed a mobile magnetic observatory that would exclude all fields but the Earth's.

Gauss's geomagnetic atlas was published in 1836. Measurements of Earth's field have continued since Gauss's first efforts, and we now have a 150-year record. One of the key findings is that the magnetic North Pole moves. It was first pinpointed by explorers in 1831, then again in 1904. During that interval it had moved 50 kilometres (31 miles). During the 20th century, the pole has moved north at around 10 kilometres (6.2 miles) per year, though that movement appears to be speeding up. It is currently moving at around 40 kilometres (25 miles) per year.

That's not the only change: records show that at mid-latitudes, compass needles are drifting about 1 degree per decade. There really is a blip in the South Atlantic too: satellite measurements tell us that under the Atlantic Ocean, west of South Africa, field lines appear to converge, forming a magnetic pole. This 'South Atlantic anomaly' exhibits its own reversed magnetic field lines, which now cover much of South America, and confuse our overall view of the Earth's field. Then there's the issue of the general weakening of the field. Taken as a whole, Earth's magnetic field has lost 10 per cent of its strength since Gauss's measurements began. To understand what that means for the future, scientists have tried to uncover the roots of the field.

Churning spheres

The fact that the Earth has a North and a South Pole might tempt us to think that its field arises from something like a bar magnet buried deep in the Earth. Unfortunately, things are nowhere near as simple as that. The Earth's magnetic field arises from a churning sphere of molten iron and nickel deep in the heart of the planet. This, the Earth's inner core, is a solid ball of iron 1,250 kilometres (775 miles) across. It is fiercely hot, at thousands of degrees, and only the pressure bearing down on it from the weight of the rest of the planet keeps it from melting.

Surrounding the inner core is the molten metal that creates the magnetic field. Heat from the inner core courses through this liquid, creating convection currents that move hot liquid metal up towards the mantle, the layer beneath the crust. This hot liquid cools as it rises, and so falls back down. The motion of this metallic conductor creates electricity, which is always accompanied by a magnetic field. The combination creates a self-sustaining 'geodynamo' that maintains Earth's field.

This geodynamo creates an extremely complex magnetic field. As the Earth spins on its axis, the field lines become twisted, creating new currents within the liquid outer core. This creates new magnetic field lines, and a new magnetic field can sometimes grow within the core. Typically, this adds to the current field, but if its orientation is shifted relative to the dominant field, it can sometimes detract from the overall magnetization of the Earth.

This may be what is happening with the South Atlantic anomaly, and it may be what is causing the apparent weakening of the Earth's field. However, researchers can't be sure, because the dynamics of the field created by such an enormous geodynamo are too complex to yield up their secrets to mathematical models. Frustrated geodynamo researchers are supplementing their mathematical models by creating their own real-world geodynamos. Typically, this involves some highly perilous equipment. If you want molten spinning metal in your lab, you can't use a metal that melts at thousands of degrees. The best candidate is something like sodium, which melts at temperatures of just under 100 celsius.

That said, sodium has its own dangers. It can burn in a fierce explosion on contact with water or air, for example. Nevertheless researchers have succeeded in spinning balls of molten sodium to simulate what is going on beneath our feet. The results have been impressive: self-sustaining magnetic fields do indeed form, and they do exhibit the kind of complex behaviour seen in the Earth's geodynamo. They even exhibit occasional 'reversals', when the North and South Poles swap places. During that process, the magnetic field fades and becomes much more complex, then grows again, but with a reversed polarity.

For a period of time, during a reversal there is no clearly

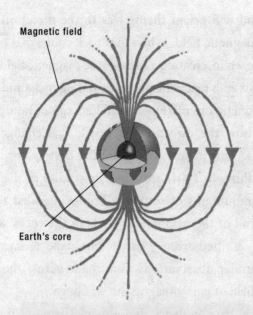

Magnetic field

Earth's core

Where the Earth's magnetic field comes from

defined field. So, could this happen with the Earth's magnetic field, with potentially disastrous results? Unfortunately, even these simulations have not yet proved accurate enough for us to make forecasts for the Earth's field. The best we can do, it seems, is to look at the evidence frozen into the planet's rocky crust, and try to extrapolate our findings.

Written in the rocks

In the molten rock that pours from volcanoes and the gaps between tectonic plates in mid-ocean ridges, magnetic crystals – tiny grains of magnetite, for example – are free to move, and will orient themselves to the direction of the Earth's magnetic field. When that rock cools, that orientation is frozen in, creating a rock whose magnetic field points towards its era's magnetic north. By dating rocks and noting their magnetic orientation, researchers can build up a picture of how the direction of 'north' has changed over millennia. This is how we gained the first evidence for a shield failure. In 1904, geomagnetic studies of the Massif Central mountains of southern France revealed that the orientation of the magnetic crystals in the rocks was significantly shifted from what it would be today. In the 1920s, similar observations were made across the world, and the field of palaeomagnetism was born.

We now have evidence that, during the past 20 million years, the Earth's field has collapsed and reversed more than 60 times. These reversals have occurred every half-million years or so, and can take thousands of years to complete. However, it is by no means a clockwork phenomenon. Sometimes, as happened during the age of the dinosaurs, no flips happen for tens of millions of years. We haven't seen a reversal for 780,000 years now. So, does that mean we are due one? Is that why the Earth's field is currently fading at what seems like an alarmingly fast rate?

We know, thanks to the logbooks kept during Captain Cook's journeys in the South Seas, that the current failure only started relatively recently. We have mariners' logs that date back to 1590, which record, amongst many other things, the direction of the Earth's field and the angle at which the field lines enter the Earth. It was a useful navigational trick – in many ways the sailors' lives depended on it. We have recorded a decline in field strength since Gauss began measuring it in 1840, but the ship's logs show no change between the 1590 value and Gauss's field strength.

Of course, it may be that we haven't enough data to draw any firm conclusions; the 'South Atlantic anomaly' may be leading us astray, for example. So should these strange measurements and discoveries give us cause for concern? Given the crucial role Earth's magnetic field has played – and continues to play – in the development of life on Earth, the answer has to be yes.

Many of the animal kingdom's most ambitious migrations involve navigation by the Earth's magnetic field. The 8,000-mile trek of the loggerhead turtle, the great American journey of the monarch butterfly, and the continent-crossing osprey all involve sensing the magnetic field. It's still not clear exactly how they do it, but we are gathering clues. The tissues of many animal species – frogs, bees, yellowfin tuna and bacteria, for example – contain the mineral magnetite, which aligns to an external magnetic field.

Migrating birds such as the bobolink have magnetite in cells in their brains. But birds have also been shown to have 'magnetic sight'. The visual neurons of migratory garden warblers contain proteins called cryptochromes, which seem to be sensitive to weak magnetic fields. When exposed to fields of different orientation, the proteins produce different combinations of chemicals. The 'blue' light of evening seems to be particularly good at stimulating the proteins to produce these chemicals, which corresponds to the time of day when birds are orienting themselves.

It's not only migratory animals that sense magnetic fields; cows are thought to be magnetically sensitive too. Satellite images of grazing dairy and beef herds, taken over six different continents, seem to show that they stand oriented to within 5 degrees of the north–south line. There are question marks over the data; it could be to do with prevailing winds, for example. Nevertheless, it is an intriguing observation, and the data seems to tie in with the various shifts between geographic and magnetic north. In Oregon, where there is a strong field, cows face 17.5 degrees off geographic north, towards magnetic north. Deer herds have been observed to do the same. So, if so many animals have this sense, what about humans?

There is no evidence we have a conscious sense of magnetic fields, but there are studies that link human health issues with magnetic fields. Research in Russia, Australia and South Africa has found links between periods of geomagnetic activity and increased suicide and depression rates. The root cause remains a mystery, but researchers have suggested that geomagnetic variations might affect melatonin production and circadian rhythms – both of which have been linked to mood disorders.

The great protector

The blue-green planet we call home sits roughly 93 million miles from the sun. We are in the 'Goldilocks Zone', where life can thrive in certain regions because the climate is not too hot and not too cold. But the sun produces more than heat. The surface of the sun is a turbulent mass of plasma, a gas composed of charged, high-energy particles. The sun is constantly losing these particles, which travel through space as the 'solar wind'. Our magnetic field directs most of these around Earth. Crucially, only a small proportion of the particles reach the Earth's surface – except, that is, when the sun creates what is known as a 'solar storm'. This often coincides with the appearance of sunspots, which are indicators of hugely intense magnetic fields beneath the sun's surface. The chaotic motion of particles means these magnetic fields writhe around, twisting and turning, and occasionally creating a whiplash that throws out a massive ball of plasma. If flung towards the Earth, its intense magnetism interacts with our magnetic field, the magnetosphere.

Depending on the relative orientation of the two fields, two things can happen. If the fields have the same alignment, they slip round one another. In the worst-case scenario, when the field of a particularly energetic plasma ball opposes the Earth's field, things get much more dramatic: the magnetic field of the plasma ball opens up a hole in the Earth's field, and the particles surge through. The

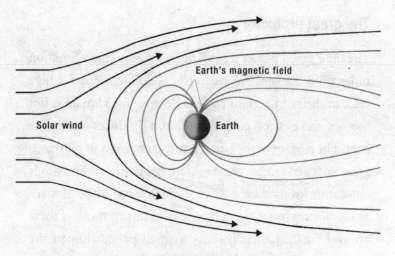

The Earth's magnetic shield

result can be devastating, damaging satellites, and causing havoc to the Earth's power grids. In March 1989, for example, such a solar storm blacked out a huge swathe of the Canadian province of Quebec, leaving 6 million people without electricity for nine hours.

When particles from the solar wind hit the atmosphere at higher latitudes, some create a cascade of energized particles. That energy is released as the fluorescent, shimmering glow of the *aurora borealis*, the Northern Lights. Some of the particles, though, reach the surface of Earth as radiation. This radiation from the solar wind has been, in some ways, a positive force. It may, for instance, be responsible for directing some of life's evolution on Earth. This radiation can damage DNA, which imposes mutations in the genetics of terrestrial life, facilitating the process of evolution.

However, the radiation is also a danger. If it is too intense, mutations in DNA can lead to sterility, cancer – even, possibly, extinction for some species. The fact that this radiation hasn't wiped out life on Earth is largely due to the fact that our magnetic field deflects most of the solar wind. So what would happen if it failed?

If the shield should fail

We know that our magnetic field came into existence at least 3.2 billion years ago. The earliest known life forms were in existence 3.5 billion years ago. The implication seems obvious: life has evolved in a magnetic field, and may require it. The moon and Mars both had magnetic fields around 4 billion years ago, but neither body has one now – nor do they harbour life, as far as we know.

Physicists' best guess about the reason for that is that their small size meant they cooled quickly, losing the heat necessary to keep a liquid core churning. Earth's larger size maintains the heat in its core, while its tectonic plates cool the mantle relative to the core. That temperature difference keeps the convection currents strong, stirring the iron-rich molten rock and maintaining our field.

And here is another connection to life: Earth's field maintains our atmosphere. The magnetic field's deflection of the solar wind means that the atmosphere is not buffeted by the solar wind's particles. Maps of Mars's little remaining ionosphere show that it is thickest where Martian rocks have maintained their magnetism. It seems that, if you lose your magnetic field, the atmosphere goes with it. So, Earth's magnetosphere does not just protect us from radiation. It also allows our atmosphere to form and develop, giving us oxygen to breathe. Are we about to lose the very air we breathe?

The answer is almost certainly not. The reversal is most likely happening, but all our experiments and observations seem to indicate that any magnetic reversal will take a few thousand years, at the very least. During this process the Earth's field will weaken, and become massively more complex, but it will remain strong enough to hold on to our atmosphere. It may not be a disaster in other ways, either.

The humans living on Earth at that time will almost certainly be at risk of receiving much more solar radiation. But no one yet knows whether that will really prove a problem. It is possible there could be mass extinction because of DNA damage, but there are so many other factors in play over those kinds of timescales that anything is possible. The last reversal didn't wipe out our ancestors, and by then we might have developed the technology to create our own artificial radiation shield. Earth's natural shield might well be failing, but this time we are ready, willing and able to face the consequences.

Why does $E = mc^2$?
The equation that underpins the universe

Go on, think of an equation. Given all those years of education, whether you enjoyed or endured them, you might think that an equation you learned at school would come to mind. But it doesn't. Instead, this one, which you probably learned by accident, pops into your head.

It is the most famous equation in the world. It appeared on the cover of *Time* magazine in 1946, and has since become part of our culture, inspiring artists and musicians, writers and film-makers. It litters the globe: you'll find it on the logo of a Japanese graphics company, a public relations company in rural England and a Toronto hair salon. Why? Because this equation summarizes how the modern world took its form.

Though the equation, written down by Einstein in 1905, was forty years old before the world saw what it could do, we shuddered at the discovery. On the cover of *Time*, it is written into a mushroom cloud looming over a fire-struck Pacific atoll. $E = mc^2$ is the equation behind the atomic bomb. It ended the Second World War, and ushered in the age of

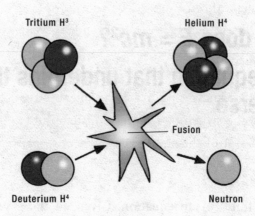

Nuclear fusion

nuclear power and nuclear threat. With it came the Cold War and, for the first time, the spectre of total destruction for the human race. Even now, with the Cold War over, the possibility that the wrong person may learn how to convert a tiny mass into an enormous amount of energy hangs heavy over us.

The happier truth, though, is that $E = mc^2$ is so much more powerful than a bomb. It is the very root of our life, our continued existence and perhaps our future too. It describes the fundamental nature of reality, revealing just how deep the illusion of the familiar notion of matter goes. If there is only one equation in your head, at least it is the right one.

So where did this equation come from? To be strictly accurate, not, at first, from Einstein. In the paper suggesting this relation between mass and energy, Einstein didn't actually write down $E = mc^2$. He wrote down $L = mv^2$, where L is the 'living' energy, m is mass, and v is velocity. It was seven years later, in 1912, that he began to routinely use E for energy, and c for *celeritas*, the Latin for 'swiftness' and a universally acknowledged symbol for the speed of light. Even ignoring the switch of symbols, Einstein didn't pluck the equation out of the air. The seeds of $E = mc^2$ were sown in laws of physics that were first formulated in the 17th century, then debated for almost two centuries.

Winding up the universe

The word 'energy' has a long history, but we have only recently begun to use it in connection with what we mean by energy today. The *Encyclopaedia Britannica* of 1842, for example, defined energy as 'a term of Greek origin, signifying the power, virtue, or efficacy of a thing'. That Greek origin, which lies with Aristotle, is actually somewhat closer to the mark. Aristotle defined energy as the source of every thing's being and function. 'Energeia', he said, was what allowed something to do its job.

In Isaac Newton's day, though, energy was still poorly defined. The concept was there: things that moved – an arrow fired from a bow, for example – had energy. When that arrow landed, however, the energy seemed to be lost. The same happened if two people collided in the street, knocking each other to the floor. Their energies, according to Newton, cancelled each other out. Before the collision there was energy; after, there was none.

Fortunately for us – according to Newton, at least – God was there. Newton felt that God, as a living and immanent deity, must be at work somewhere in the universe. One of the deity's vital roles, Newton suggested, was to top up the cosmic energy reserves. God was there to wind up the clockwork universe and keep the planets moving through the heavens, but he also applied himself to everyday situations – to colliding peasants, for example.

It was not a view shared by Newton's great rival, the atheist Gottfried Leibniz. In an acerbic comment on Newton's view, Leibniz said he found it hard to understand that God Almighty would have to wind up his own watch from time to time. 'He had not, it seems, sufficient foresight to make it a perpetual motion,' Leibniz wrote in a 1715 letter to the philosopher Samuel Clarke. Newton and Leibniz were already rivals over the authorship of the mathematical tool known as calculus, which had enabled Newton to calculate the orbital motions of the planets. This conflict over energy, too, could be boiled down to another mathematical issue.

Newton had formulated the energy of a moving body as mv, the product of its mass m and velocity v. Leibniz, on the other hand, thought it to be mv^2, the product of the mass and the square of its velocity. The difference had a profound effect. In Newton's formulation, two identical bodies moving in the opposite direction with the same velocity would have energies mv and $-mv$. If they collided, the resulting energy would be zero. Leibniz's squaring of the velocity meant that the 'negative' direction made no difference, because a negative quantity always squares to a positive number. In Leibniz's formulation, the energy was not lost from the universe.

For a number of years, the question was simply a matter of ideology. If you were English-speaking, you liked Newton's work and ideas, you thought of energy as mv. If you spoke German, you sided with Leibniz, and squared the velocity. This jingoism was overcome through a Dutch and French collaboration. Willem 's Gravesande, a Dutch scientist, had been dropping weights into soft clay from various heights. The depth of the hole the weights made was, presumably, proportional to the energy, which in turn must be proportional to the height from which they were dropped and the speed on impact. The only way the sums worked was if energy was indeed proportional to the square of the velocity. 's Gravesande didn't see this for himself. It was a French noblewoman called Emilie du Châtelet who put all the pieces of the puzzle together in the first half of the 18th century and declared Leibniz the winner. The energy due to motion – living, or kinetic energy – was proportional to the square of the velocity. E depends on a velocity squared.

Though 's Gravesende and (in particular) du Châtelet had made great strides forward in illuminating the relationship between a body's motion and its energy, they still had no idea what happened to all the energy once the motion had stopped. Did it disappear? The answer to that question only came after the discovery of a principle called 'conservation'.

Conservation work

The first experimental hints at a general principle of conservation came in the late 18th century. In an astonishingly careful set of experiments, carried out just a few years before he was guillotined at the behest of the Paris revolutionaries, the French scientist Antoine Lavoisier monitored how a variety of materials changed with burning, rusting, or some other natural process of change. He found their mass was always conserved in some manner.

Each of the experiments was carried out in a closed container, and the substance under investigation (together with any air or water in the chamber) was weighed before and after the experiment. Within the limits of his experiment, the total mass of material in the chamber remained constant. Even something as violent as combustion, which altered the physical form of a material so radically, still did not push materials out of existence. The mass measurements told Lavoisier it remained there in the experimental chamber; altered in form but still there nonetheless. Things didn't simply disappear from the universe, but they could be transformed between different forms.

That probably comes as no surprise to you. Thanks to a couple of centuries of experiments such as Lavoisier's, we have come to accept that the universe is, effectively, a 'closed' system, containing a finite amount of 'stuff' that can be transformed from one state into another. And the most fundamental transformable – but always conserved – quantity is energy.

After taking thousands of years to get to grips with the concept of energy, it still took almost the whole of the 19th century for scientists to work out that energy is always conserved in nature. With hindsight, it seems a little odd that this revelation was so slow in coming. It had long been known that kinetic energy could be converted to heat. Those who bored the barrels of cannons, for example, knew that the process generated vast amounts of heat. But it was only with the invention of thermodynamics, the branch of science that relates temperature and heat to the motion of atoms and molecules (see *Why is There No Such Thing as a Free Lunch?*), that we discovered exactly how that worked.

The heat revolution

If $E = mc^2$ rules the modern world, thermodynamics created it. The discovery that heat was a form of energy, and thus could be converted to kinetic energy that would perform work, was revolutionary in every sense. Heat a bath of water sufficiently, and its conversion to steam – when under pressure – could move a piston. And moving a piston could change the fate of nations. The discovery of machines such as the engine and the refrigerator, powered by mechanical work or heat, created the Industrial Revolution, the foundation of the modern, technological age.

This conversion of heat energy into kinetic energy is just one example of how energy is conserved, moving between multiple forms but never disappearing from the universe. 's Gravesande's weights, for example, had 'gravitational potential energy' before he dropped them. That potential energy came from the energy, stored in his muscles, used to lift them to the drop height. That energy came from the food he ate, and that, in turn, came from his food's ultimate energy source: sunlight. When the weights hit the clay, their potential energy, ultimately derived from sunlight, was converted to kinetic energy (or movement) in the clay, some heat energy (due to friction), and sound energy. The energy did not disappear from the universe.

Similarly, a paraffin lamp contains potential energy. When the paraffin burns, the stored chemical potential energy is released as heat and light. The heat energy will be imparted to the molecules in the air around the lamp, and will manifest as kinetic energy: the molecules will move faster.

What seems surprising, though, is that energy can take the form of mass. Mass is surely very different to energy: while mass is associated with solidity, energy seems transient and ephemeral. But there is a link – and it is found in James Clerk Maxwell's equations of electromagnetism.

Maxwell's embrace

In the 1830s, Michael Faraday showed how electricity and magnetism are inter-related: electricity produces magnetism, and vice-versa. Shortly afterwards, Maxwell came up with a series of equations that detailed exactly how this process worked. Many physicists looking into Maxwell's equations saw that they contained the essence of mass. It was well known, for example, that a box containing electromagnetic fields weighed more than one containing none. The question was, what did that mean?

The mainstream view was that inertial mass – the resistance of a body to movement – lies in the fact that charged particles would be difficult to move in the vicinity of their own electromagnetic fields. True to his character, Einstein did not follow the mainstream. Instead, he found the answer in one of the flaws of the equations.

Maxwell once said that his equations describe a 'mutual embrace' between electricity and magnetism. However, it is actually a three-way embrace: electricity and magnetism do not exist without movement; the motion of charged particles creates electricity and magnetism. And herein lay a deep problem. Analysis of experiments showed that motion could invalidate the equations. If the emitter of electromagnetic radiation was moving relative to the observer, the equations no longer predicted the correct values for the electromagnetic field.

This was what motivated Einstein's 1905 paper 'On the Electrodynamics of Moving Bodies', in which he introduced special relativity. Einstein's genius was to insist that the laws of physics be consistent however you are moving through space. To do this, he modified Maxwell's equations so that you couldn't move in any way that changed the speed of light from its absolute value, c. The speed of light is an unassailable constant. Move towards a stationary source of light, and the light will always come at you with speed c. Move away, and you will measure it passing you at c. And here is where we find the link between energy and mass.

Light carries mass

Einstein suggested that the existence of energy – any kind of energy – brings with it an associated mass. As he stated in a letter to his close friend Conrad Habicht, shortly after publication of the $E = mc^2$ paper, 'The relativity principle, in association with Maxwell's fundamental equations, requires that the mass be a direct measure of the energy contained in a body; light carries mass with it.'

The first implication Einstein noted was for radioactivity: if radium was giving out energy, it ought also to lose some mass. The German physicist Max Planck saw a more prosaic (but in some ways more profound) implication. A hot object – a frying pan, say, will weigh more than a cold one. This was a revolutionary idea – even today it still seems strange.

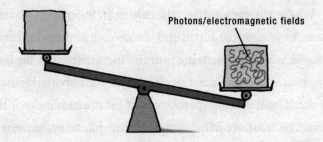

Photons/electromagnetic fields

The mass of radiation

Nevertheless, it is absolutely correct. We now have good evidence that mass is just one way of carrying energy. You can move, and carry kinetic energy, but you can also lock your energy in by simply existing. To see why, we need to explore the origin of mass.

Where mass comes from

You are made of particles that, at their root, have no independent mass. They get their mass from a quantum phenomenon known as the 'Heisenberg uncertainty principle'. At root, this says that every quantity in nature has a fuzziness; it doesn't have a fixed value. That's true even of the energy of empty space: while we think of it as having zero energy, it is actually fizzing with energy that manifests as pairs of 'virtual' particles that disappear as quickly as they appear. These fleeting, ghostly particles, it turns out, give the frying pan its mass.

When you shrink down in scale from frying pan, to iron atom to iron nucleus, you end up looking at particles called quarks, which make up the protons and neutrons in the iron nucleus. When physicists work out the mass of quarks, there is nowhere near enough to account for the heaviness of the pan. The mass actually exists in virtual particles that manifest from the fuzzy energy of empty space. Experiments involving high energy particle collisions and the crunching of millions of numbers have confirmed that these 'gluons' act to hold the quarks together in the proton and neutron, and that the energy involved is what we see as most of the mass of the pan.

Hence the hot pan weighing more. Given that almost the entire mass of a frying pan comes from the fizzing energy of empty space, it doesn't seem quite so hard to believe that adding a bit more energy, in the form of heat, also adds to the mass. The ability of high-energy processes to release this energy is what lies at the heart of our existence. When hydrogen atoms fuse in the sun, eventually forming a helium atom, the process releases some of their gluon energy (energy that we call mass) as heat and light – the very heat and light that created life on Earth.

Thanks to the colossal size of the speed of light, coupled with the fact that $E = mc^2$, there is a surprising amount of energy locked up in ordinary matter. A single walnut, for example, has enough potential energy locked within it to power a city. We have released something like this, of course

– not with walnuts, but with atoms of uranium. Suitably prepared, their gluon energy can be released to provide electrical power to cities – or to bomb them.

Whether in bombs or in power stations, we have measured the mass of the particles we begin with, the mass at the end of the process, and the amount of energy released. In every case it has been shown to be true: E really does equal mc^2. The most accurate proof we have of the validity of Einstein's equation was carried out in 2005. Unsurprisingly, it involved painfully sensitive measurements. The energy measurement for the left hand side of the equation, for example, required a team of researchers to measure the energy of a gamma ray photon to around one part in 1 million.

For the mass side of the equation, meanwhile, the researchers had to measure how the mass of an ion changes when it gives off a gamma ray photon. That is a tiny change in mass, equivalent to seeing a hair's breadth change in the distance from New York to Los Angeles. There were no nasty surprises: the researchers found a startling agreement between the two measurements. It seems that E really does equal mc^2, to better than one part in 2 million. You can rest easy: that one equation you know is pretty solid.

Can I change the universe with a single glance?

Spooky quantum links and the chance to rewrite history

Einstein put this question a slightly different way. In the early 1950s, he once turned to the young physicist Abraham Pais, raised his eyebrows, and asked, 'Do you really believe that the moon only exists when you are looking at it?'

Einstein had spent the last two decades growing increasingly frustrated by the pioneers of quantum theory. Their ringleader, Niels Bohr, claimed that the weirdness inherent in the theory, such as atoms existing in two places at once, or effects preceding their cause, could only be explained if nothing – not even the moon – really existed until it was measured or observed.

Einstein's question to Pais was an exasperated appeal to common sense. The idea that something as big and as permanent as the moon could be at the mercy of a tiny human observer hundreds of thousands of miles away is nonsensical. But that doesn't necessarily mean it is nonsense. With the advent of quantum theory at the beginning of the 20th century, the ridiculous had become the sublime. Pais remembers wondering why Einstein was so stuck in the past. 'Why does this man, who contributed so incomparably much to the creation of modern physics, remain so attached to the 19th century view of causality?' he wrote in *Subtle Is the Lord*, his biography of Einstein.

Even in the 18th century, people had been questioning the nature of reality. Bishop George Berkeley famously asserted that, if no one was around to hear a tree falling in a forest, the tree would make no noise. What's more, the tree doesn't even exist unless someone is perceiving it. Fortunately, Berkeley suggested, our common sense is preserved because God is always present to act as observer.

Niels Bohr took the same approach to the quantum world: the only proper interpretation of the vagaries of quantum theory, Bohr said, is that nothing has any properties or existence until it is observed in some way. Einstein's refusal to accept this idea isolated him from the development of quantum theory. What's more, his best attempt to refute it ended with its confirmation. Every experiment we have carried out suggests that, yes, you can change the universe with a glance. The means of your power? A quantum phenomenon known as 'entanglement'.

Tangled up in space and time

Erwin Schrödinger called entanglement the defining trait of quantum theory. He first spotted it in 1935, noting that the equations of quantum theory, applied to two interacting particles, impart an unusual quality. After their encounter, they can no longer be properly described as individuals. They are linked; the quantum description of particle A – its momentum or spin, for example – contains information about particle B, and vice-versa.

That has a very strange consequence. If you change the properties of particle B, you necessarily change the properties of particle A. This doesn't require a physical link; the entanglement link changes properties whatever the separation between the two entangled particles. Two suitably prepared entangled particles can instantaneously change each other's quantum state even when at opposite ends of the universe.

Einstein was having none of this, and dubbed it *spukhafte Fernwirkungen*: 'spooky action at a distance'. It showed, he said, that there were still gaps in quantum theory. And, with the help of two friends called Boris Podolsky and Nathan Rosen, he set out to prove it. The scenario the trio outlined is still the gold standard for proving the weirdness of the quantum world. It is known as the EPR paradox, and concerns the fate of two pairs of particles, with each pair separated from the other by an enormous distance.

Bell and the spooky action

The most rigorous experimental version of the EPR paradox

was drawn up in 1964 by John Bell of the European Organisation for Nuclear Research (CERN), the particle physics laboratory based in Geneva, Switzerland. Bell imagined separating two entangled electrons, and sending them to experimenters on opposite sides of the Earth. The experimenters then simultaneously measure the electron spin. The details of the set-up are complex, but Bell's challenge was that, if orthodox quantum theory was right and Einstein was wrong, particular kinds of measurements would show a correlation between the two spins.

Einstein died before he could see Bell's experiment performed. The first implementation was by the French physicist Alain Aspect in 1982, but there have been innumerable tests since, and they all confirm Einstein was indeed wrong. Entanglement is indeed a spooky action at a distance, one that denies the objective existence of anything. Bell's electrons take on their properties only when a measurement is performed, that is, only when someone looks at it.

Even if you already believed that a tree falling in a forest makes no noise, it is still truly remarkable to note that, to stretch the analogy, cutting the trunk of one tree can fell another – even when they are in separate forests. A pair of entangled electrons affect each other instantaneously, and from across the universe. It really is as spooky as Einstein claimed: the standard interpretation of time and space seems to wither to nothing in the light of quantum entanglement.

Beam me up . . .

Entanglement is already being put to use. Quantum cryptography, for example, makes use of the 'remote control' function, combined with the fact that entanglement connections are extremely fragile, to give a means of securing information. It is rather like the historical practice of placing a seal on important communications; messages encrypted with entanglement are tamper-proof because any attempt to eavesdrop breaks the connection.

Perhaps more conceptually impressive (but of less practical use) is the quantum version of teleportation that entanglement allows. It is a complex operation, but the basics are that a measurement on one of an entangled pair of particles forces a change in the properties of the other. Done with suitable skill and subtlety, the distant particle can be imbued with all the characteristics of the original without ever being in the same place. It involves other particles too, and the transmission of some information through 'normal' channels, so it is, perhaps, more of a telefax than a teleportation. However, it is still an impressive innovation. Although it can only be done on single particles, such as photons, so far, there is no fundamental reason why we can't extend the technology to transmit more and more quantum objects – perhaps an atom or more.

This will undoubtedly prove useful: although it is unlikely that we will ever achieve *Star Trek*-style human teleportation, moving quantum states around in this way promises to allow information processing on an unprecedented scale. Many research groups around the world are trying to develop 'quantum computers' that will perform computations at speeds that are exponentially greater than anything a normal computer can manage. Quantum state teleportation will play a key role in the way these machines work. Our role in shaping the universe needn't stop with observations that determine the existence and properties of a few particles in quantum experiments, however. According to the late John Wheeler, one of the most respected scientists of the 20th century, each of us can change the very history of the cosmos.

A great smoky dragon

Wheeler's assertion arose from considerations of quantum measurement. It is now widely accepted that one of the strangest manifestations of quantum theory is the phenomenon where, given the option, something like a photon of light will take all available paths. This 'superposition' results from the wave-like character of quantum objects. A single photon fired at a screen scored with two narrow slits will produce a pattern of dark and light bands on a screen positioned on the other side of the slits. This is an 'interference pattern', and is associated with wave behaviour. To produce interference, however, the photon must have passed through both slits. How could a single photon do that?

It seems reasonable that we should be able to resolve this by watching the photon. If we look at the slits, we will find which one the photon went through. But any attempt to determine which way the photon went destroys the interference pattern. In this scenario, the photon behaves like a bullet, passing through one slit or the other, and produces no interference pattern.

Bizarrely, the photon seems to behave like a wave when nobody is looking, and a particle when they are. The idea of a photon making some kind of conscious choice related to things in its environment is ridiculous to physicists' ears. This is why Einstein and others said we must be missing something; there must be some 'hidden variables' that determine the photon's behaviour.

Wheeler suggested a way to test this. What if, he said, we only looked at the photon's path after the photon had made its 'choice' of how to behave? Would that alter the photon's behaviour? Wheeler's 'delayed choice' experiment is not an easy one to perform, but physicists have managed to do it. In experimental set-ups where a photon takes just 14.5 nanoseconds to traverse the apparatus, researchers have managed to change the set-up after the photon has made its 'choice' of whether to behave like a wave or a particle. Nine nanoseconds after the photon entered the apparatus, when the photon has already split like a wave to go through two slits or, like a bullet, gone straight for one slit, the researchers attached a detector to one of the slits.

What was the result? With a detector in place, there was no interference. With no detector, there was interference. This is exactly what standard quantum theory predicted: the presence of a detector forces the photon to behave as a particle, and particles don't interfere. If the weird behaviour could be explained by the existence of hidden variables, the photon would already have been 'committed' to one behaviour or the other. Before experimenters chose whether or not to detect the photon, it would have manifested as a wave or a particle, with no option to save the choice till after it had passed through the slits. Wheeler referred to the result as revealing quantum processes to be a 'great smoky dragon'. Its tail – the input – we can know. Its mouth – the outcome – is also clear. But the body of the dragon is a cloud of impenetrable smoke, and 'we have no right to speak about what is present,' Wheeler said.

What's more, Wheeler added, we can say the same about the processes of the universe. The emission of light from stars is a quantum process, after all: the individual photons of starlight have much the same character as the photons of laser light we use in quantum experiments. And a delayed choice experiment carried out on a cosmic scale is the same as one done in the laboratory – but with much deeper implications.

Changing cosmic history

In a provocative thought experiment, Wheeler used the phenomenon of gravitational lensing to illustrate his point. When light from a distant quasar travels towards Earth, it may pass close to a huge galaxy. The galaxy's mass bends the light, giving us the illusion of two quasars where there is only one. Einstein cited this phenomenon as a prediction of general relativity, and the prediction was borne out when the British astronomer Arthur Eddington measured the effect in 1919.

In Wheeler's view, this lensing is just a double slit experiment on a grand scale. A photon coming from the star has two paths it can take. If we had a way of observing interference effects as a result of the two paths, we would see an interference pattern. Wheeler's object of choice was the

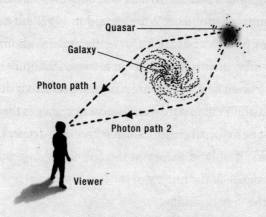

The delayed choice experiment

quasar 0957+561A,B. It is just over 7 billion light years away, and thanks to a galaxy that sits between us and the quasar, we see two images of it in our telescopes. The quasar's light takes 7 billion years to reach us, then, and a good portion of that journey is after the lensing galaxy. So, Wheeler said, we can take our time, and think about whether we want to measure it with a particle detector or a wave detector. Whatever we decide will have determined whether that photon took one path around the galaxy, or both.

And we can make that decision billions of years after the photon has passed the galaxy in question. 'This is the sense in which, in a loose way of speaking, we decide what the photon shall have done after it has already done it,' Wheeler wrote in 1981. Bohr's idea that nothing has properties until a measurement has been done seems odd. But, with his cosmic thought experiment, Wheeler had out-absurded him. Suddenly, Bohr's idea seems a lot less difficult to swallow.

Quantum phenomena, Wheeler said in 1992, 'are neither waves nor particles but are intrinsically undefined until the moment they are measured. So, in a sense, Bishop Berkeley was right when he asserted "to be is to be perceived". But we could equally well turn it around and assert that the fact the delayed-choice experiment has been run successfully in the laboratory, if only in theory in the cosmos, shows that we can participate in the history of the universe – perhaps right back to the very beginning.

'How did the universe come into being? Is that some

strange, far-off process beyond hope of analysis? Or is the mechanism that came into play one which all the time shows itself?' This quote finds Wheeler wondering if the Big Bang was a quantum event – the universe pulled into existence by something that also governs everyday life.

His 'participatory' universe, where the role of observers changes the quantum nature of cosmic history, goes some way to arguing for this. But the Cambridge University cosmologist Stephen Hawking goes even further. The quantum world, he says, allows us to determine the whole history of the universe – from where we are right now. He calls it 'top-down cosmology', and thinks it solves that perennial question: what came before the Big Bang?

What came first?

Many physicists would say that trying to discuss things prior to the Big Bang is as ridiculous as asking what lies north of the North Pole. After all, time came into existence with the Big Bang (see *What is Time?*), and until the clock of time is ticking, we cannot consider concepts of 'before' and 'after'.

Hawking, though, is not willing to stop at the moment of creation. It is, he says, quite reasonable to ask what kicked the universe into being. It is a subtle and difficult argument, as one might expect, but it draws upon well-established ideas in physics. The first of these is an interpretation of quantum theory put forward by Wheeler's most famous graduate student, Richard Feynman.

It is called the 'sum over histories', and suggests that quantum processes follow all possible paths simultaneously. In the double-slit experiment, for example, the interference pattern comes from the photon going not just through both slits, but through every other possible path, such as bouncing off the surface of the moon before it hits the detector. All the various paths have an associated probability, which comes in positive and negative flavours, rather like a wave. When everything is added together, the sum describes what we tend to observe in an experiment.

When Hawking applies the sum over histories idea to the universe, he really does mean histories. It bears some relation to an enormous 'what-if?' experiment, pulling together every possible scenario for the story of the universe. In one scenario, our solar system failed to form. In another, gravity is wildly increased. Events change too: Hawking has to consider a history in which Elvis is still alive, for example. Each of these scenarios has an associated probability.

Your ever-changing universe

What's even more disturbing than all these strange possibilities is the fact that they are subject to the measurements we are making today. Just as choosing to measure for a particle or a wave changes the outcome of a quantum double-slit experiment, Hawking reckons the way we gaze out at the universe today can change the way it evolved billions of years ago.

Hawking admits it is a strange idea – but it only seems strange because we are inside the universe in question, he says. Someone looking from outside the universe would see nothing strange, in Hawking's view. And such an observer could see how the universe came into being from nothing. That is possible because Feynman's sum over histories – and thus Hawking's calculations – rely on a notion of 'imaginary time'.

Although it sounds fantastical to use imaginary time, it is not as much of a stretch as it sounds. Engineers routinely use numbers composed of real and imaginary components to describe and predict the behaviour of electrical circuits.

In Hawking's top-down cosmology, the sum over histories of the universe, calculated using imaginary time, changes normal time into a spatial dimension. The result of this is that the problematic 'beginning' of the universe disappears. Back when the energy of the universe was packed into the tiniest of volumes, everything ran according to quantum rules, and what we call time was actually a dimension of space.

So, according to this interpretation of quantum theory at least, time emerged from a change in the nature of space. Hawking's flexible, changeable universe is, on some levels, extremely attractive. It gives us participation in the universe, and does away with the problem of the pre-Big Bang state. But it is far from being universally accepted as the answer we have been looking for.

The important point to keep in mind is that no one understands how the quantum world really works. The interpretation you follow, whether it is Feynman's sum over histories or Bohr's rejection of an objective reality, is in many ways just a matter of taste. That's why most physicists consider themselves to be adherents of the 'shut up and calculate' interpretation of quantum theory. In this pragmatic position, no one knows what it all means, but we do enjoy playing with it – and that can be enough.

The Cornell University physicist David Mermin put it best. The predictive power of the theory 'is so beautiful and so powerful that it can, in itself, acquire the persuasive character of a complete explanation'. But it is not an explanation. So, while quantum theory says you can change the universe with a single glance, remember that we still see quantum theory through an obscuring fog of ignorance and incomprehension. It's OK to believe you can make the moon appear. Just don't try to convince anyone else of your powers.

Does chaos theory spell disaster?

The butterfly effect's influence on weather, climate and the motions of the planets

It's not a particularly new idea. You might even have grown up with the concept; it is written into a familiar children's rhyme:

> *For want of a nail, the shoe was lost*
> *For want of a shoe, the horse was lost*
> *For want of a horse, the rider was lost*
> *For want of a rider, the battle was lost*
> *For want of a battle, the kingdom was lost*

That is chaos theory, sometimes known as the 'butterfly effect'. Can the lack of a single nail destabilize geopolitics? Can the flap of a butterfly's wings result in a storm thousands of miles away? The answer is yes, and it happens all the time. Not necessarily in those exact ways, of course. The children's rhyme is obviously a playful look at consequences. And even the butterfly was born in a throwaway comment.

Edward Lorenz, who initiated research in this field, was due to give a talk at the 1972 meeting of the American Association for the Advancement of Science, but had failed to provide a title. The meteorologist Philip Merilees, the session chair, eventually came up with something. In a paper nine years earlier, Lorenz had mentioned how a meteorologist had scorned chaos theory, pointing out that if it were correct, 'one flap of a seagull's wings could change the course of weather forever'. Merilees evidently remembered this line, and concocted a variant that has entered popular culture like few other scientific concepts. The title of Lorenz's talk was 'Does the flap of a butterfly's wings in Brazil set off a tornado in Texas?'

The official term for the butterfly effect is 'sensitive dependence on initial conditions'. The basic idea is that most systems that change over time – whether they are natural, such as weather, or artificial, such as the numerical output from a computer program – will turn out very differently if even the tiniest adjustment is made to their starting point. This simple observation has such profound consequences that it has given rise to a whole new field of research.

The repercussions of chaos theory, as this field is known, have been felt across the whole of science. From the dynamics of the planets to the way epidemics spread through human populations, chaos theory's influence is as wide-reaching as it is important. The entire universe is, it seems, in a state of chaos. This discovery would have come as a terrible shock to the grandly titled Pierre Simon, Marquis de Laplace. In the 18th century he had embraced the Newtonian revolution with relish. His book on the mechanisms of the universe, where he took Newton's gravitational theory and used it to map out the movements of all the planets, was a masterpiece. A few years later, he boldly boasted of science's power to tame every known phenomenon:

An intelligence which at a given instant knew all the forces acting in nature and the position of every object in the universe — if endowed with a brain sufficiently vast to make all necessary calculations — could describe with a single formula the motions of the largest astronomical bodies and those of the smallest atoms. To such an intelligence, nothing would be uncertain; the future, like the past, would be an open book.

Within a few decades of Laplace's death, however, that vision had begun to unravel. It started in 1860, when the Scottish physicist James Clerk Maxwell discussed the amplification of small changes when considering what happens when molecules collide. Thirty years later, Henri Poincaré discovered that the mutual gravitational attraction of three moving objects displayed sensitive dependence on initial conditions. Then, in the 1920s, the Dutch engineer Balthasar van der Pol found chaos in the tones produced by a telephone handset connected to a vacuum tube. The electric current driving the tube occasionally triggered what we would recognize as feedback, and van der Pol was able to write down an equation to describe it.

Though this equation was enormously useful to engineers trying to build vacuum tubes into electronic systems such as broadcasting equipment, the whistling itself was no more than a nuisance. In fact, various mathematicians and engineers studied the phenomenon without noting anything particularly remarkable about it. Though the story of chaos theory has involved many players, it was Edward Lorenz who really brought it to life.

A short cut to chaos

Lorenz had been a weather-watcher since childhood, and spent the Second World War as a weather forecaster for the Army Air Corps. Some years later, when working as a researcher at the Massachusetts Institute of Technology, Lorenz combined meteorology with mathematics and the relatively new science of computing. He built a processor that could model a simple version of the weather. And it was here that he discovered the butterfly effect.

As with many of the most important breakthroughs in science, it happened by accident. One afternoon in 1961 Lorenz was short of time and halfway through a weather simulation on his computer. Looking through a printout of where he wanted to start, he punched in the numbers that would rerun the simulation from halfway onwards. It came out wrong – or at least wildly different from the original.

Alarmed by the apparent error, Lorenz checked what he had used as an input. He had, he noted, cut the numbers off after the third decimal place, assuming that the fine detail would make no difference. Where the computer had been using 0.506127, Lorenz had input only 0.506. It had made all the difference in the world. Lorenz had discovered the sensitive dependence on initial conditions: an unpredictability that arises through our limited knowledge.

Our rulers are not infinitely small, our actions are not infinitely smooth, our machines are not infinitely powerful. Thus every measurement we perform, and every computation carried out using those measurements, will have a small but finite error. Before Lorenz, that might have been considered to cause a problem as small as the error. But sensitive dependence on initial conditions means that, more often than not, the error will eventually be huge.

Everywhere we look we find chaos. The solar system, for instance, is chaotic because it involves the interaction of more than two bodies. As Henri Poincaré proved, while there are solutions to the equations that describe interactions between two bodies, add one more – or many more – and no exact solutions can be found. The mathematical equations describing the system simply cannot be solved.

With eight planets and a sun to consider, not to mention countless rocks, asteroids and comets, chaos reigns in the heavens. But a solar system that runs by chaos rather than clockwork does not mean we are in danger of colliding with another planet at any moment. Chaotic orbits are often 'bounded', moving in cycles that never quite repeat, but within a limited space, thus limiting the danger of collision.

Strange beauty

This boundedness, where chaos operates within strict limits, has given rise to another icon of chaos: the 'strange attractor'. Imagine a simple system that exhibits chaotic behaviour, something like a double pendulum, where two rigid rods are loosely joined and allowed to swing freely. The free movement of the double pendulum is similar to the movement of your leg from the hip, but with the knee able to bend in two

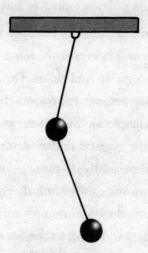

The double pendulum

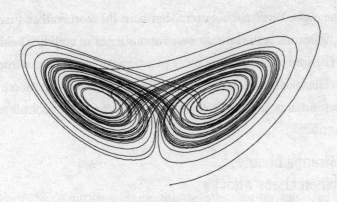

The Lorenz attractor

directions. Until you have seen it with your own eyes, it is almost impossible to fathom the degree of unpredictability the double pendulum exhibits. It swings back and forth, with the endpoint of each oscillation seemingly at random.

Swinging freely from some arbitrary starting point, the bottom of the double pendulum traces out a pattern composed of loops that form definite shapes. Though it never makes the same loop twice, it also doesn't deviate far from the established pattern. This 'attraction' to a particular form is what gave the strange attractor its name. Perhaps the most famous example is the butterfly-shaped Lorenz attractor. This is a map of the movement of a chaotic system in three dimensions. As the motion continues, the lines are denser and denser. But the trajectory never crosses itself, and never repeats.

A similar pattern emerges from a steel pendulum set in motion just above three magnets laid out as the corners of a triangle. Each magnet exerts a pull on the steel bob, and the pull varies as the bob moves in and out of the magnet's field. The sum of all the competing pulls forces the pendulum into a chaotic trajectory that is sensitive to the tiniest variation in its initial position or velocity. This is illustrated in its strange attractor.

When chaos attacks

The fact that these chaotic orbits are bounded does not mean they can't have extreme consequences. This is demonstrated in the effects that the planets can have on each other. Though the orbits themselves do not stray too far from their expected paths, the chaotic motions might occasionally create a cataclysmic threat. Calculations show that a tiny kick to Saturn, from the particles that compose the solar wind, for example, can turn its orbit 'aperiodic'. That means it will take a slightly different path each time it goes round the sun.

It's a scary prospect, because it opens up the possibility that Jupiter, Saturn and the sun will align at some moment. The combined gravitational pull of this trinity is enough to pull rocks out of the asteroid belt that lies between the orbits of Mars and Jupiter, and could unleash an asteroid storm. There have been claims that such an event preceded the asteroid impact that seems to have ended the age of the dinosaurs. If that is what happened, it would not be the first time that chaos has had an impact on biology – and it was certainly not the last. The butterfly effect rules biology as surely as it rules the swing of a double pendulum.

Nature's chaos

Chaos theory has had an enormous impact on the science of ecology. The idea of populations growing in good times and thinning out in the bad times has always been a strong part of biological thinking, but the rise of chaos theory and the butterfly effect created a Eureka moment. Before chaos came along, biologists of a mathematical bent would write down equations that approximated to the situations they were interested in. They might describe, for instance, how many squirrels were living in a square mile, how rich their food source was, how frequently they reproduced, and how many predators shared the territory.

From that, they could work out how that population of squirrels would grow and diminish over time. But any time the equations gave results that seemed to be going out of control, the mathematical biologists would 'reset' the scenario, assuming that there was some instability in the system that needed to be reined in. With the advent of chaos theory, it became clear that the wild changes could quite easily be a natural part of the system.

Imagine, for instance, a population of squirrels with no predators. If, on average, each adult produces less than one offspring per generation, the population will dwindle to zero.

If the number of offspring is between one and three, there is some stability. If the average number of offspring per generation is more than three, however, things get strange.

A tendency to 'boom and bust' appears in the population statistics. It is, essentially, the same as that whistling valve tube in the telephone line mentioned earlier: a process of feedback creates wild oscillations. Chaos theorists call it a 'bifurcation'. It means that the population is uniquely sensitive to the number of offspring per generation. In one year, the population will boom, but in the next it busts. There's nothing in between. Then the bifurcation splits again and again, and things eventually start to look random: there is no apparent pattern. But only for a while: as time goes on, the number of offspring increases again, and, out of nowhere, another bifurcation comes into play.

Such complexity is everywhere in the natural world, and understanding it can save lives. The hit and miss, up and down, boom and bust occurs in epidemics of diseases like AIDS, measles and polio, for instance. Because the number of cases follows a chaotic trajectory, it is sensitive to a knock from something like an inoculation programme. But that knock doesn't always wipe out the disease; instead, the numbers can be thrown into an unstable regime – around a bifurcation region, for instance. This means that short-term figures for the disease might rise, suggesting that a programme of inoculation has failed. Awareness of chaos allows medical researchers to see beyond the initial response, and allow for the chaotic response, mapping what is hopefully a downward trajectory over the long term.

An understanding of biological chaos and the butterfly effect is saving lives around the world in a more immediate way too. Your heart beats because of co-ordinated pulses of electricity that work through the cells in a kind of wave, causing the muscle to contract in specific ways and at specific times. When this breaks down, an 'arrhythmia' occurs. Heart arrhythmias kill hundreds of thousands of people each year: for myriad reasons, the heart can stop beating normally – or, indeed, at all. Often the muscles are all contracting randomly, and the heart is no longer a pump, but a seething, pulsating mess of tissue. It's a chaotic system – and one where a good kick can take the chaos away.

Medics have long known that a jolt of electricity can set this problem right again, but you can't just put any old jolt into a human heart. To set the rhythm right again requires an understanding of its chaotic dynamics. The heart is, effectively, an oscillator like a pendulum. And when you know how a chaotic pendulum can be controlled, you can also design a defibrillator that works much better than the ones designed by trial and error. The main area where the butterfly effect has been put to work, though, is exactly where it started: the weather.

Predictably unpredictable

Meteorologists like to run hugely complex simulations of the Earth's weather systems on massive supercomputers. The simulations are based on the laws of physics, and model things such as how ocean and air currents move around the globe. Before the butterfly era, we might have thought that a suitably powerful simulation would predict the weather weeks, months or maybe even years in advance. Chaos tells us this is just not possible.

The trouble is, the physics of the models is approximate, and the data used to set up the simulation even more so. The weather stations used to gather information are scattered over the Earth, with large gaps between them: we don't have the information from the places between the weather stations. Scientists now know that, within just a few days, these sources of error are enough to set the weather models off on a trajectory that will bear no relation to what actually happens to the weather. A butterfly flapping its wings somewhere between the weather stations could cause a storm that no one saw coming.

Of course, the meteorologists reset their models every time they get new data in. They also run 'ensemble' forecasts, where they put slightly different initial conditions into the model and look at how much the outcome varies. That allows them to create an averaged forecast that is likely to be more accurate than any one prediction. It also allows a measure of the reliability of their forecast. It's not enough to have a prediction – it's better to have an idea of how much you should trust that prediction.

Ironically, the longest-term predictions come out OK: the science of climate prediction is not so sensitive to initial conditions as are short-term weather predictions. That's essentially because climate prediction deals more in generalities than specifics. The flap of a butterfly's wings might cause a storm in Texas, but another flap might calm a storm that was already blowing up. Over the 30-year average that constitutes a climate analysis, the number of storms evens out, and each butterfly becomes irrelevant.

Lorenz used the equations of chaos theory to show this. When you look at a strange attractor, you see a particular shape. Applied to climate science, the shape you see indicates the future climate. The flowing line that moves around unpredictably, gradually creating the shape, is like noise on the signal; it is not the parameter of interest. That means running climate simulations, however chaotic their predictions might be over the short term, reveals a reliable broad-brush picture of what is coming. So, does chaos theory spell disaster? Quite the opposite: that flapping butterfly has been instrumental in warning us about the greatest threat faced by humanity: runaway climate change induced by human activity. Chaos is not always a problem.

What is light?
A strange kind of wave, and an even stranger kind of particle

'What is poetry? Why, Sir, it is much easier to say what it is not. We all know what light is, but it is not easy to tell what it is.' Samuel Johnson thought this a convincing justification for the difficulties of defining poetry. Unfortunately, the idea that we all know what light is has one fundamental flaw. It's not clear that we do.

When Johnson wrote those words in 18th-century England, Isaac Newton's view of light as particles or 'corpuscles' of energy reigned supreme. Within 20 years of Johnson's death, Thomas Young had 'proved' light was a wave, not a particle. A century later, Albert Einstein showed light to be, once again, particles. Now we have to think of it as both – or neither. Light, the universal metaphor for understanding and revelation, is astonishingly opaque.

One thing about light is certain: it is essential to our existence. Without light from the sun, plants would not be able to use photosynthesis to harvest energy and grow, and we would have nothing to eat. Humans deprived of light suffer depression – researchers who kept rats in the dark for six weeks watched their brain cells die for lack of light. Insufficient exposure to direct sunlight creates skeletal problems such as rickets. Whatever light is, we need it.

This was recognized by ancient civilizations. The Neolithic monument at Stonehenge is, it seems, a temple to the light-giving sun. The Egyptians worshipped Ra, the sun god, as the giver of life. The first people to attempt a definition of light, though, were the ancient Greeks who were a little more circumspect than the Egyptians: to them, light was not something to be revered, but a by-product of fire, one of the four fundamental elements that made up the universe.

There were various Greek ideas about the nature of light and vision. Euclid's was the most developed. Light from an object mixed with the light from the eye, he said, but a person could only see the object when the eye's fire reflected directly back from the object. It was only close to the modern scientific view in that it had light travelling in straight lines, however. And we had to wait for nearly two millennia before anyone attempted to move our understanding of light forward. That progress was kicked off by a Frenchman, René Descartes, early in the 17th century.

From waves to particles to waves

Not that Descartes's contribution lasted. His idea was that space is filled with an invisible fluid that he called the 'plenum'. The plenum, Descartes said, has a 'tendency to motion' such that a candle creates a pressure in the plenum in much the same way as a drum creates sound waves in the air. This pressure is passed on to the eyeball and manifests as light. Almost as soon as he started thinking about it, Isaac Newton debunked the idea.

If light is just pressure of the plenum on the eye, Newton argued, then breaking into a run on a dark night should flood the world with light. Newton was a big fan of the emerging idea of the atom – that, on the smallest scales, everything can be divided into component parts. Light, he argued, should be no different. It was, he suggested, composed of atomic elements that Newton referred to as 'corpuscles'.

The corpuscular theory reigned for 150 years, but it didn't get an easy ride. Newton's great rival Robert Hooke had produced a competing wave theory (the wave theories of this time assumed the existence of an 'ether' in which the light created vibrations), and so had the Dutch mathematician and astronomer Christian Huygens. Both ideas passed muster in experiments; it was really only Newton's reputation that gave corpuscles their sticking power. Then, in 1803, Thomas Young performed the definitive demonstration of the wave nature of light.

Young's demonstration centred on the fact that the interaction of two water waves produces predictable geometric patterns (see *What Happened to Schrödinger's Cat?*). Where the crests of the waves meet, there is a 'constructive interference': a crest that is double the size. Where two troughs meet, 'destructive interference' makes the trough twice as deep. Where a crest meets a trough, flat water results. Knowing the speed and direction of travel of the water waves, and their initial separation, it was possible to predict the resulting wave patterns in the water. If light is a wave, the same phenomenon should arise when light is passed through a double slit. The two interacting light waves should produce an 'interference pattern'.

Interfering with the ether

Young's double slit experiment, now a staple of the high school science laboratory, worked beautifully. It killed the corpuscular theory outright: light was unquestionably wave-like. Only one question remained: if light is a wave, what does the wave move through? The original answer was similar to Descartes's plenum: the ether, a ghostly substance that filled space and time and provided the medium through which electricity, light and magnetism were transmitted. However, an experiment performed at the end of the 19th century showed the ether did not exist, at least not in any way that enabled the transmission of light.

In 1887, Albert Michelson and Edward Morley set out to show the ether *did* exist. Their interferometer experiment comprised a rotating table that would measure light's speed in varying directions. The idea was that, if an 'ether wind' was blowing, light would move at different speeds in different directions. This difference would show up in the interferometer, shifting the interference pattern around.

The experiment failed to detect an ether, a fact that stunned physicists at the time. Even though it was clear that light could travel through a vacuum, and so was fundamentally different from sound, it was assumed that it still needed to travel in something. Light showed wave characteristics, but if there was no ether for it to travel through, light was not the kind of wave anyone had encountered before. We have arrived at a conundrum that remains unexplained. Yes, light is a wave. But it is unlike any other wave. And some of the best minds in physics insist that it is not a wave at all.

The corpuscle rides again

Perhaps Richard Feynman put it most forcefully. 'I want to emphasize that light comes in this form – particles,' he once told his students. 'It is very important to know that light behaves like particles, especially for those of you who have gone to school, where you were probably told about light behaving like waves. I'm telling you the way it does behave – like particles.'

If Feynman's insistence is not enough reason for you to give up on the wave picture, consider this. Albert Einstein proved that light comes in particle form. His experiment, which was published in 1905, was called the 'photoelectric effect', and its physics lies behind the workings of solar power. It had been known for some time that light hitting the surface of a metal could release electrons from the metal. No one understood, however, why the flow of released electrons seemed to increase as the frequency of the light moved towards the high-frequency, ultraviolet end of the spectrum of light. Common sense – as dictated by Maxwell's electromagnetic theory – said that the current should increase with the intensity of the light, not its frequency.

Einstein solved the puzzle with the notion of a photon: a packet of energy that was the quantum particle of light. In Einstein's prediction, the number of electrons released from the metal would depend on the energy of the photon – which is proportional to the frequency of the light 'wave'. Only photons with a certain minimum of energy would be able to free an electron. Photons hitting the metal with energy above that threshold would not only free an electron, but would impart their extra energy to it. Experiments that measured the kinetic energy of emitted electrons showed this to be the case, and Einstein was awarded the 1921 Nobel Prize in physics.

It is perhaps unfortunate that the greatest physicist of the 20th century, the creator of general and special relativity, should get his Nobel Prize for the discovery of the photon. Despite the Nobel Prize and despite Feynman's insistence, the idea of particles of light remains one of the shakiest concepts in physics. It is easy to think of photons as particles in

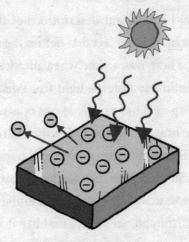

The photoelectric effect

the same way that we think of electrons or protons as particles. But photons are much less particle-like than that. They don't have mass, for instance.

The physicist Willis Lamb, who made many important discoveries during a stellar career at the famous Bell Labs, went so far as to claim the word 'photon' should be banned from physics. At the very best, he said, the word should be licensed – with Lamb giving licenses out only in cases where he felt there was a real need to move away from the wave-like picture of light.

Lighting up the cosmos

If we can't quite put our finger on the fundamentals of the nature of light, there is still a lot we can say about the primacy of light's role in our descriptions of the universe. Most important of all is the fact that light is the fastest thing in the cosmos. There was a time when physicists believed light travelled infinitely fast, the light from distant stars or planets appearing instantaneously at our eyes whenever we looked up at the heavens. By the end of the 17th century this idea had died, as experiments showed that a finite speed of light could explain anomalies such as the irregular orbit of Io, Jupiter's innermost moon. The idea that nothing could travel faster than light, though, seemed at first like it was plucked from nowhere.

This notion arose from considerations of the Maxwell equations that describe electromagnetic phenomena. We know that the motion of an electric charge – electricity – causes a magnetic field to appear in its vicinity. That same magnetic field generates electricity as it grows. And so the cycle repeats. Maxwell found that this resulted in something that moved with the undulating intensity of a wave, and that he could calculate the speed at which it moved forward. It was a well-known value, the same speed that astronomers had found (by measuring the timings of eclipses and the orbits of planets and moons) to be the speed of light. Therefore, Maxwell reasoned, light must be an electromagnetic phenomenon.

That was all very well until physicists came to appreciate that not all electromagnetic phenomena were well behaved. Analyse the radiation being emitted by something moving relative to you, and you'll find that it doesn't obey Maxwell's equations exactly. Einstein fixed this with a radical step. Presuming that the laws of physics should be the same however anyone is moving, he made a new law: that the speed of light is always constant – and nothing can move faster than this.

Fixing light speed

When a car's headlamps move past you, the light they are emitting is not speeded up by the motion of the car. What's more, if the car then slows down, the light does not slow down. It is always travelling at just under 300 million metres per second. This counter-intuitive notion, which lies at the core of Einstein's special theory of relativity, has enormously strange consequences (see *What is Time?*). But it has been shown to be true in countless experiments. And, according to Einstein's relativity, the closer you get to the speed of light, the harder it gets to accelerate further.

This effect restores order to the universe, allowing Maxwell's equations to describe any situation, no matter how much relative motion there is between the emitter and detector of the radiation. Bringing things right up to date, we now have a quantum version of Maxwell's work, known as quantum electrodynamics, or QED, and this describes light's behaviour perfectly. Whatever light is, the realization that it travels at a constant speed irrespective of circumstance has enabled us to map the past, present and future of the cosmos.

When we see light coming from a distant star, we know that it has travelled through time as well as space. Our view of the sun is always as it was eight minutes ago; light from other stars gives a view much deeper into the past. What's more, by seeing stars in their various stages of development, while knowing how far away they are, we can tell what will happen during a star's lifetime – information that we can use to predict what will happen in the future. The path our sun will follow, for example, is now well understood: we have about 5 billion years before it begins to die, a process that will see it bloat into a 'red giant' and engulf most of the planets – including Earth.

A mysterious power

The other great application that has arisen from our deeper grasp of light is the technology that perhaps best defines the 20th century: the laser. In an age of CD players, supermarket checkout scanners, high speed optical phone cables and corrective eye surgery, it is hard to believe the inventors of the laser didn't know what it might be used for.

Laser is an acronym, standing for Light Amplification by the Stimulated Emission of Radiation. The light from a standard bulb, or even the sun, comes from atoms that release light on an individual basis. The principle behind the laser is to pump energy into a gas of atoms and then release it in a controlled way: the pulses are 'coherent', which means they are locked together to give an intense, powerful beam.

This can be achieved by priming the atoms with a jolt of energy that kicks one of their electrons into a high energy state. A second jolt knocks that electron down again. It emits a photon of light as it falls, kicking off a chain reaction. Each photon kicks another one out of its higher energy state, stimulating the emission of more photons from other atoms in the gas, which in turn stimulates more emission. The result is a laser beam.

It is easiest to explain the mechanism behind lasers using the photon approach, but the power of the beam is best suited to a wave description. We know that waves, such as water waves, can add up if they meet 'in phase', that is, with their peaks coinciding. The result is an enormously powerful wave, which is, essentially, what laser light is like.

It is not just the power of lasers that makes them so useful. The fact that the light is so tightly controlled, with the photons locked together, makes them a great scientific tool, used in myriad ways from finding the distance to the moon to probing the secrets of the atom. The spin-off applications, such as scanning barcodes, reading information from CDs and making the modern telecommunications industry possible, are just the icing on the laser cake.

Strangely, lasers even make it possible for you to go faster than light. In 1999, Lene Hau slowed light down to the speed of a passing bicycle. This is possible through the use of two

lasers: one to 'prepare' some sodium atoms, and one to provide a pulse of light for 'slowing'. The energy of the
preparation beam is tuned to a value where it pushes the
sodium atoms into a state where they cannot absorb the
'slowing light pulse'. That means the pulse, which would normally be absorbed, travels through the cloud of atoms.

As it travels, though, it gives some of its energy to the
sodium atoms. These atoms hold the energy for a moment,
then release it into the travelling pulse again. The result is
rather like plucking the front carriages off a train, and
reattaching them at the back. While the carriages are moving
at a normal speed, the progress of the train as a whole
is impaired.

So, although light's speed is reduced, it's really a sleight of
hand. Two years later, Hau went a step further and stunned
the world by stopping light in its tracks. To do this, the
quantum state of the atoms is altered even further, to the
point where they hold the energy for as long as is required.
By nudging the atoms out of that state, Hau could release
the light.

Nothing about the tricks we can do with lasers resolves the
debate over whether light is a wave or a particle, however.
For all our ability to interpret what light tells us about the
universe, and use its power to change the world around us,
the nature of light itself remains elusive. Thomas Young
would be astonished to discover that we can now mimic his
double slit experiment – the experiment that proved light is

a wave – and tweak the set-up so that the only viable explanation is that light has a particle nature. We still don't have a good way to tie together the wave and particle view. This puzzle is at the heart of quantum theory, and remains a mystery (see *What Happened to Schrödinger's Cat?*). What is light? It's a wavicle.

Is string theory really about strings?

The vibrations that create our universe

Well, no. It's actually about a universe composed of elastic loops that are formed from stretchy strands of energy joined together. But, as at least one physicist has put it, as a name rubber-band theory lacks dignity, and a little dignity seems appropriate for a theory that is our best hope of finally understanding the universe.

String theory might be billed as a new, ultramodern idea, but it is not. It first appeared in 1968 as a result of our post-war infatuation with particle physics. We only discovered the atomic nucleus in 1911. We learned to split the atom in 1938, and within 20 years had learned almost everything there is to know about nuclear physics. Ten years after that, string theory, an audacious attempt to broaden these new horizons to encompass the whole universe, was born.

It arose because an Italian physicist called Gabriele Veneziano spent his youth poring over the results of experiments that smashed protons together at high energies. Eventually, he began to see a pattern in the data: two colliding protons would cause particular kinds of particles to shoot out from the crash site at predictable angles. The initial products of these collisions were quarks, the particles that make up protons. However, the quarks subsequently combined to give different kinds of particles. Most of these particles were unstable and short-lived. When Veneziano drew physicists' attention to the predictability of their formation, a few of them began to piece together an explanation. The results made sense, they said, if you forget about the idea that particles are tiny points of matter, and pictured them instead as lengths of string. The energy they carry makes them vibrate, and as the particles gain and lose energy, these strings lengthen and shorten. As the vibrating strings collide, the resulting range of vibrations give rise to what we interpret as different kinds of subatomic particles. It seems impressive now, but nobody saw this as a triumphant 'theory of everything' straight away. In fact, one of the originators of string theory had his paper rejected as rather insignificant. But then it did have significant problems.

String vibrations

Teething problems

In nature, particles can be (roughly) split into two types. The 'fermions', such as the electron, or the quark, make up the matter. The 'bosons', such as the photon, are the particles that transmit forces. String theory set the rules for bosons, but had nothing to say about the existence or behaviour of fermions. Since fermions account for the basic constituents of matter, that was a big flaw. But it wasn't the only one.

If physicists were to take string theory seriously, they had to make the theory consistent with the twin pillars of physics in the 20th century: quantum mechanics and relativity. The only way the string theorists could manage that was to envisage a universe that contained 25 dimensions of space while also admitting the existence of particles that could never be brought to rest and of others that travelled faster than light.

It was all a bit much to swallow, and, for a few years, string theory lay neglected and unexploited. It didn't help that string theory was meant to be a means of describing what is known as the strong interaction of nuclear physics, the force that holds quarks together to form protons and neutrons. While string theory was lying fallow, what is now known as the 'standard model' of particle physics emerged. This tied together everything we knew about the subatomic particles in a very neat package. String theory looked superfluous, if not a little foolish.

So how did string theory become the answer to physicists' prayers? Most of the hard work was carried out in 1970 by a French physicist called Pierre Ramond. He found the string vibrations that gave rise to fermions. As a bonus, this also removed the need for faster than light particles and reduced the number of required extra dimensions to just nine. String theory now became 'superstring theory', and – hallelujah – it was consistent with quantum theory and relativity. There was just one more flaw to address: the string theorists' suggestion that some particles could never stop moving.

The solution to this turned out to be the one that really mattered. String theory's meteoric rise to fame came about from the discovery that its unstoppable particles were ones that physicists had long been hoping to create in a fundamental theory: the photon, the quantum particle of light, and, most excitingly, the graviton: the quantum particle of gravity.

Where gravity comes from

If it was good news to find a theoretical justification for the photon, then the discovery that string theory gave the graviton was the stuff of physicists' dreams. Since the inception of quantum theory in the 1930s, physicists have wanted to find where gravity and the other forces meet. Here, perhaps, was the answer.

The various forces of nature – the strong and weak forces

that act in the nucleus, and the electromagnetic force that acts between charged particles – seem to have a fundamentally different nature to the gravitational force. Gravity plays by different rules. It only attracts, for instance, where electromagnetism attracts and repels. Ultimately, physicists were aiming to explain this uniqueness. And string theory seemed to be able to do just that.

In string theory, the ends of the strings are associated with a particle and its antiparticle – an electron and a positron, say. The vibration of the string carries the force that acts between this pair of charges. A string can break into two, or collide with another string. The result of all this produces strings that occasionally close up into a loop. There is no charge associated with this loop of string, only a force that matches the characteristics of the force we know as gravity.

The realization that string theory has gravity built in turned on lightbulbs over countless physicists' heads. All the while they had been looking at string theory as a means of describing nuclear interactions, but what they actually had was a quantum theory of gravity, a grand unified theory, a theory of everything. Almost overnight, string theory became the great new hope in physics. That hope has long been deferred, however. It was 1984 by the time string theory seemed poised to complete the task Einstein had started. And that is more than three decades ago. So where is the promised final theory? That, it turns out, is a very contentious question.

A final theory?

Nobody doubts we need a final theory. Quantum mechanics and relativity are inconsistent with one another – almost to an absurd degree. The laws of physics, for example, are different for quantum particles moving in different ways through the universe. The quantum description of an electron at rest is different from the description when the electron is moving at something near the speed of light. Albert Einstein had constructed the theory of relativity precisely to avoid such problems.

Looking at it from the other direction, relativity doesn't make sense when viewed through a quantum lens. Quantum calculations can be done without reference to time or distance, for instance, but relativity can't cope with anything that doesn't need time or space. Potentially, string theory can overcome all these problems. But it hasn't yet – at least not to the point where it is overwhelmingly appealing as a theory. Before it can be hailed as the new über-theory, before we can say that the universe is indeed made up of strings, the theory has to conquer its own demons. One of those was immediately obvious from the start. We live in just three dimensions of space, but to be consistent with relativity and quantum theory, string theory initially needed 25. That was later reduced to just nine – but that's still six dimensions that haven't ever been seen. So where are they?

The short answer is that, from our perspective, at least, the

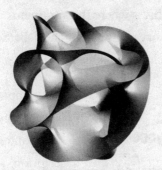

A Calabi–Yau space

extra dimensions are rolled up very small, or 'compactified'. Imagine a hosepipe seen from a distance. It looks like a one-dimensional line rather than the three-dimensional object it really is. String theorists say that this is how we must think of the extra dimensions. They are there, but hardly make any impact on our three spatial dimensions.

This is not just a hand-waving argument, but has been made with mathematics: the extra dimensions can be rolled up into six-dimensional torus, or any one of more than a million complex six-dimensional shapes known as Calabi–Yau spaces. This, naturally, brings a lot of flexibility to string theory. Each of these Calabi–Yau shapes, for instance, come with a set of variables to specify its exact nature. These will give different characteristics to the compactified dimensions, and have knock-on effects in the dimensions that are open to view.

The result of this is that string theory does not describe one universe, but many, all with slightly different properties. String theory, therefore, creates myriad universes of every shape and size – and this is where the arguments about the usefulness of string theory really begin. The debate is simple: do we see this multiplicity of possible worlds as a problem or an opportunity?

A universe for everyone

String theory's critics call it a theory of *anything* rather than a theory of everything. Anything goes, they say: unless someone finds a way to sift our universe from among the possibilities, string theory can make no falsifiable predictions about the nature of our universe. Thus, can we really call it a true scientific theory? Many string theorists reject this criticism outright. If string theory gives us so many universes, their argument goes, perhaps that's because there *are* so many universes.

There is certainly something to this argument. Modern cosmology tells us that the universe most likely went through a period of rapid 'inflation' just after the Big Bang. Put simply, the universe blew up like a balloon, its size increasing by a factor of 10^{30} – that is, it got 1,000 billion billion billion times bigger – in just a fraction of a millisecond.

No one knows why this should have happened, but it is the best explanation for some otherwise puzzling features of the cosmos. The universe is homogeneous: it seems the same everywhere. This is odd because the Big Bang would have created it otherwise. But the mystery can be solved by inflation: the universe that goes through a period of rapid expansion early in its life will become homogeneous.

Inflation also happens to be a useful support for string theory. If the inflation happened once, there is no reason why it wouldn't have happened again. Every patch of space–time is subject to the same laws, and so every patch can, in theory, generate a bubble universe that grows until it pinches off and floats away. So countless other universes will bubble out from every universe, blowing up and pinching off to an independent existence. Each one will have slightly different properties from the others. The laws of physics, in other words, will be different – there could be a universe without gravity, for instance, or one where there are 17 different kinds of electron. In this view, the universe is actually a multiverse: a landscape of universes that take every possible form. Somewhere among these universes is ours.

It has to be said, there is no experimental evidence to support this idea, just a deductive argument: that inflation, though devoid of explicit experimental support of its own, is the best explanation for the features of our universe, and thus might be applicable again and again. Worse still, there can be no evidence – at least in the scientific terms of falsification put forward by the philosopher Karl Popper.

Is it science?

The standard idea in science is that you make a hypothesis and see if experiments can falsify it. Hypotheses that withstand attempts at falsification gain support, and can eventually be developed into theories. The string theory idea of the landscape of universes cannot be falsified in these terms. There is no way to make any prediction about the properties of our universe compared to another – the other universes are simply not accessible to our experiments.

It is possible to make a virtue of this; string theorists have made much of the observation that the expansion of our universe is accelerating, for example. There is no good explanation for why this should be, and the string theorists have jumped on the lack of explanation as a kind of backhanded proof: maybe there is no explanation, they say: maybe it is an example of how our universe is just one possibility. In other universes, the laws of physics work to keep the expansion constant, and yet others have a decelerating expansion. Diversity is the only law. Whether this makes string theory a white elephant to science is an ongoing debate amongst physicists. But the fact remains that we don't have a better way forward at the moment.

There are other attempts at building a theory of everything. Perhaps the most advanced is 'loop quantum gravity' or LQG. This suggests that space is ultimately composed of indivisible quanta that are around 10^{-35} metres in size. A network of links between these quantum nodes – imagine an airline route map – creates the space–time we live in. The particles that come together to create our familiar world of atoms and molecules are created when quantum fluctuations induce knots and tangles in this space–time.

Or that's the idea. LQG is not yet a well-defined answer to the problem of unifying quantum theory and relativity. In fact, there are probably only a hundred or so researchers working on it worldwide. Which means string theory, with its workforce of thousands, is maintaining its dominance. Eventually, though, the plan is to replace it with another theory: M-theory.

What lies beneath

Rather surprisingly, no one is sure what the M stands for. Whatever its true provenance, however, the M of M-theory has come to be associated with membranes. To make the mathematics work, string theorists have postulated that the 11 dimensions of string theory are populated by surfaces called 'branes' (short for membranes) as well as the strings. These branes can have up to nine dimensions.

Though they add to the richness of string theory, wrapping around compactified dimensions, providing an anchor point for wandering strings and allowing new kinds of universe that might exist, their most famous role might be in establishing what – according to string theory – came before the Big Bang. The idea is that our universe came about through a collision between two four-dimensional branes. The enormous kinetic energy of the colliding branes creates a vast amount of heat: the Big Bang fireball and, crucially, the standard zoo of particles known to physics. This scenario is known as the 'ekpyrotic' universe, taken from the Greek phrase 'born out of fire'.

Interestingly, the ekpyrotic universe does away with the need for inflation because it is created homogeneous. Doing away with inflation undermines the idea of an infinite landscape of varied universes. And that means we don't have to give up on creating falsifiable hypotheses about why our universe is as it is. Having said all that, only a minority of string theorists subscribe to the ekpyrotic view of the universe, and it may be that only a minority of physicists have any faith in string theory's power to explain the universe. So where will this go? Can we at least test string theory? This is another contentious question. As yet, after four decades of work, we have yet to find a way to properly test the string idea. But there are some possibilities.

A peek into extra dimensions

One of the hopes has been that we will see hints of the hidden extra dimensions. Such a hint could be an anomaly in gravity as we examine its effects on ever-smaller scales. Gravity is an 'inverse square' law: double the distance between the two objects under test and the force between them drops by a factor of four. Triple the distance, and it drops by a factor of nine. But with a tiny rolled-up dimension in play, that inverse square law may not describe exactly what is going on. Gravity may work slightly differently between objects that are less than a millimetre apart, for example.

So far we have seen no evidence of this. Tests of the inverse square law down to less than six-tenths of a millimetre have shown up no such anomaly. Perhaps we shouldn't be surprised, though. Strings themselves are tiny, after all – less than a trillionth of a trillionth of the diameter of an atom. How could we detect such an incredibly small thing? One hope is that some of them have grown because of the expansion of the universe. As the cosmos has grown, some cosmic strings might have expanded into 'superstrings' that might be scattered through space. It is possible that we could detect their presence through their effect on light travelling to us across the universe: the high mass of the superstrings would bend the light as it passed, creating an optical illusion known as gravitational lensing.

Then there's the idea that in the standard, non-ekpyrotic universe scenario, inflation would have created ripples in the gravitational field of the early universe. These 'gravitational waves' should have been preserved in the cosmic microwave background (CMB) radiation, the echo of the Big Bang, but string theory places limits on how strong those ripples should be. If they were large, they would have unfurled some of the compactified dimensions, and we would have more than the three dimensions of space we currently experience. So string theorists are hoping for no gravitational waves in the CMB. Again, it's hardly a conclusive test, though. As yet, there are no 'direct hit' experiments that will give us a definitive yes or no to the theory. Is the universe made of strings? It's a definite maybe.

Why is there something rather than nothing?

The Big Bang, antimatter and the mystery of our existence

Could there be a bigger question? Why is it that we, the galaxy, the universe, everything, exist at all? To understand the answer, we have to go back to the beginning of everything – if, that is, we can find one.

In many cultures, there is no such thing as a beginning. The ancient Greeks, for example, revered the concept of the circle, and everything essential to the universe, including the universe itself, existed in ever-repeating cycles.

Until the beginning of the 20th century, the consensus amongst astronomers was very much the same: our universe had existed for all eternity and it made no sense to talk about a beginning. Which, of course, made the church authorities slightly uncomfortable.

The Book of Genesis begins with a beginning: a something was created out of nothing. Perhaps that is why a young Belgian priest called Georges Lemaître decided that astronomy needed to give consideration to a point of genesis for the universe. Lemaître, a professor of physics and an accomplished astronomer, was the first to suggest the idea that came to be known as the 'Big Bang'. His hypothesis was that everything arose from a 'primeval atom', which split apart to produce all the matter in the universe. Starting with Einstein's equations of general relativity, which describe the dimensions of the universe, he showed that its radius could change – the universe, in other words, could expand.

It was more than a theoretical consideration: there was evidence for it too. Astronomical observations gathered by Lemaître and others showed that most galaxies were moving away from ours. An intriguing implication was clear to Lemaître. Perhaps the galaxies were moving away because Einstein's space–time was expanding? Lemaître's resulting paper suggested that we live in an expanding universe, spawned by the explosion of what he called 'the cosmic egg'.

Something from nothing

The Pope was thrilled with Lemaître's work; astronomers less so. The idea that the Pope would approve of a scientific theory – that their data and theories were supporting the doctrine of creation *ex nihilo* – sat uncomfortably. Nevertheless, within a few years the English astronomer Edwin Hubble had propelled the idea of a beginning to the universe to the very forefront of cosmology. Hubble took Lemaître's work forward, gathering together data from many different astronomers, and supplementing it with his own. He showed definitively that almost all the galaxies were flying away from us at enormous speeds, and that the universe must be expanding.

The idea that the universe hadn't always been as we see it remained a subject of intense debate for decades, however. It was only in 1963 that the conclusive evidence – the cosmic microwave background radiation, sometimes known as the 'echo of the Big Bang', was found. At that point, almost all opponents of the Big Bang became convinced it was indeed our best explanation for cosmic history. With the advent of Big Bang cosmology came an answer for why there is something rather than nothing. But it was only a partial answer. The idea raises obvious questions: 'What caused the Big Bang?', and 'What banged?'

Physicists have taken diverse paths here. Some say the questions are meaningless because time came into existence at the moment of the Big Bang; the notion of 'before' therefore makes no sense. It is, they say, like asking what lies north of the North Pole. Others make some attempts at an answer, but the answers are little more than untestable speculations. They invoke quantum phenomena such as the Heisenberg uncertainty principle, which says that nothing can have an exact amount of energy, and that includes a universe with zero energy. Quantum fluctuations, then, will give rise to a universe with some amount of energy, and there are processes that can amplify that to create a Big Bang.

There are some physicists – notably Stephen Hawking – who say the Big Bang was not the beginning of everything, but resulted from processes occurring in other dimensions (see *Can I Change the Universe With a Single Glance?*). Others take this further and suggest that we are in a 'cyclic universe' that is caught in a never-ending cycle of creation and destruction as objects known as 'branes', which exist in these other dimensions, repeatedly crash together and move apart (see *Is String Theory Really About Strings?*). Such reasoning is satisfying to those who want to believe there is no need for a divine hand in creation, but unconvincing to those who don't. These issues, it seems, may lie beyond the reach of science.

But even given the problems of describing how a Big Bang arose, there is another, later issue that ought to have made sure that, very soon after the something was created, there was nothing again. Before Edwin Hubble had laid the foundations of Big Bang theory as an explanation for our existence, another Englishman, Paul Dirac, was undermining them. The issue is blown wide open thanks to Dirac's greatest contribution to physics: antimatter.

Where did all the antimatter go?

Dirac was a strange, quiet man with few social skills. One much-reported conversation sums him up rather neatly: during a formal dinner at Cambridge, Dirac sat next to the equally reticent E.M. Forster. Their entire conversation through the multiple courses consisted of one exchange. In reference to a scene in Forster's novel *A Passage to India*, Dirac asked, 'What happened in the cave?' Forster, much later in the meal, replied, 'I don't know.'

The pair evidently whiled away the hours in their heads. In Dirac's case, that was certainly productive. The existence of antimatter, now known to be an essential part of the zoo of subatomic particles, was not suggested because of an experimental result in need of a theory. It arose directly through Dirac's considerations of the governing equation of quantum theory: the Schrödinger equation (see *What Happened to Schrödinger's Cat?*).

When describing the energy of a quantum particle, Schrödinger's equation threw out something that was, at first glance, an impossible puzzle. The energy of a fast-moving particle, it said, involved two numbers. These two numbers, when multiplied together, would give a result of 0. Each one multiplied by itself, however, had to give the answer 1.

In any normal mathematics, this simply cannot be done. But using arrays of numbers known as matrices, Dirac did it. The only price to pay was that the energy of the quantum particle could be negative as well as positive. And, through a tortuous chain of reasoning, Dirac showed that the negative energy particles could manifest in our world. They would look like familiar particles, but with some strange adjustments.

In 1928, in a series of talks, Dirac proposed the existence of an antielectron. It would look exactly like an electron, but would have a positive charge. He was ridiculed: physicists of the time considered matter to be made from negatively charged electrons and positively charged protons, and nothing else (the neutron was still four years from discovery). Undaunted, Dirac published his theory three years later. The antielectron, he said, would be 'a new kind of particle, unknown to experimental physics'. When it met an electron, there would be an explosive annihilation, Dirac predicted. And the same would be true of any particle: each and every one had an antimatter nemesis.

The announcement may not have thrown down a gauntlet to the physics community, exactly – Dirac was largely uninterested in what anyone else thought. Nevertheless, the prediction was out there. And unknown to anyone, the evidence was out there too. Physicists studying cosmic rays, the charged particles that smash into Earth's atmosphere, creating a cascade of other particles, had seen – but not understood – the signature of an antielectron five years before Dirac made his pronouncement. When passed through a magnetic field, some of these particles bent the 'wrong way'. It was noted as anomalous, and discussed at scientific meetings around the same time that Dirac was discussing his theoretical ideas. But it was not until 1932 that anyone put two and two together, when Carl Anderson discovered the antielectron in the debris of cosmic ray collisions. The breakthrough won Anderson a Nobel Prize.

Antimatter in the universe

Once it was clear that antimatter could exist, it was only natural to ask how much of it there is in the universe. Is it all around? Does it sit unnoticed in antimatter stars in antimatter galaxies? And if so, is there less antimatter than matter in the universe? Would that explain why there is something rather than nothing? The trouble is, answering these questions involves understanding a lot more about antimatter. But how do you study something that annihilates on impact with everything around you?

We have found a few answers out in space and are now reasonably sure that there are no antimatter stars out there, although there are natural sources of antimatter in the universe. One, seen by the INTEGRAL telescope, is a fountain of positively charged electrons, or positrons, that streams out from somewhere near the centre of the Milky Way. There are clues on Earth too. As Carl Anderson showed, we can study antimatter by looking at the debris created when cosmic rays smash into the Earth's atmosphere. But it's not an abundant source: high-energy cosmic rays smashing into gas clouds only produces around 3 or 4 tonnes of antimatter per hour in the whole of our solar system.

In fact, our efforts to create antimatter on Earth are even more feeble. The prime source of antimatter is the European Organisation for Nuclear Research (CERN) in Geneva, but it is an incredibly crude process. The CERN researchers simply smash beams of positively charged protons into a lump of metal – copper or tungsten. The result of this is a huge spray of particles, a few of which are negatively charged anti-protons. A few of these few particles spray out in the right direction to be harvested in a trap.

For every 10 billion joules of energy the CERN researchers put into the process, they get one joule's equivalent back as antimatter. If you annihilated all of the antimatter ever made at CERN, the released energy would power a single electric lightbulb for no longer than a few minutes.

Not that you can store it up to feed into a power grid. Antimatter can't be allowed to touch normal matter, so it can only be held by the electromagnetic fields of a 'Penning trap'. This uses magnetic fields to hold the particles away from the physical walls of a container. Scientists can only store anti-matter in a Penning trap for a few minutes at a time, and each trap can only hold so many particles: as soon as their mutual repulsion overcomes the repulsion due to the trap's magnetic field, the antimatter annihilates on the walls of the container.

What's more, CERN's traps can only hold around 1,000 billion particles, which sounds like a lot, but isn't. It is around a hundredth of the number of atoms contained in a child's balloon – and that would only come from hundreds of millions of years' worth of production at CERN. The dream of antimatter-powered spaceships taking us to the stars will have to wait for a better source of fuel. But the dearth of antimatter hasn't stopped CERN giving us clues as to why something survived and nothing didn't.

Losing the balance

Experiments in CERN's Large Electron and Positron Collider (LEP) tell us that the Big Bang would have created a universe in a situation where five cubic metres of space held 10 billion antiprotons, and 10 billion and one protons. Today, that same space contains one proton and no antiprotons. At some point in our history, matter and antimatter met and annihilated one another, leaving just one proton per five

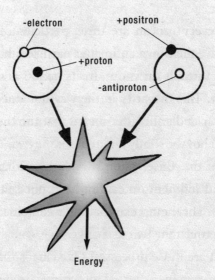

Annihilation between hydrogen and antihydrogen

cubic metres. These protons eventually came together and formed the universe we know today. So what created that initial imbalance of particles?

In the late 1960s, the Russian physicist and dissident Andrei Sakharov set his mind to solving this puzzle. The detective story is not yet finished, but we do have a clue about what created the initial imbalance between matter and antimatter. It seems to be an odd little particle called a neutrino. Sakharov's biggest clue came in 1964, when physicists found something odd about the weak force, which governs radioactive decay and various other processes that take place in the nucleus of an atom. The weak force, unlike every other force, does not act quite the same way on matter as on antimatter.

Inside every proton are three particles called quarks. Quarks have their own antimatter particles, the antiquarks. The weak force, it turns out, treats quarks and antiquarks differently. This disparity in the way the weak force deals with matter and antimatter means that the laws of physics must somehow be subtly different for the two. That, in turn, means that the conservation laws that cover things such as energy and momentum cannot be applied to matter and antimatter. There must exist natural processes that allow the balance between the two to change.

So what are those processes? One clue is that they must have occurred in a period of cosmic upheaval, where reactions were taking place between particles, antiparticles and radiation. If the reactions took place at one rate for particles, and another rate for antiparticles, you would get a net difference in the amounts that survived. The early universe, which was a long way out of thermal equilibrium and thus buzzing with transformations of energy into particles, and particles into different particles, was perfect for creating such an imbalance.

That's as far as Sakharov got, but we have now gone much further. We know that the hot, dense conditions just after the Big Bang were perfect for creating a host of particles that would never be seen in our cold, empty universe. And one of them, known as the 'majoron', is the reason why there is something rather than nothing.

Enter the majoron

As the majoron ages, theory suggests that it does not respect the symmetry laws that create equal numbers of particles and antiparticles from one decaying particle. The majoron decays to form particles called neutrinos, which are tiny uncharged particles that zip around the universe at almost the speed of light. Majorons also produce the neutrino's antiparticle, the antineutrino. But there is no compulsion for the majoron to form equal numbers of neutrinos and antineutrinos.

During their lifetime, the neutrinos and antineutrinos will collide with electrons and positrons to form quarks and antiquarks. If there is an excess of neutrinos over antineutrinos, that means more quarks than antiquarks will form. So when annihilation happens between quarks and antiquarks, there is matter left over.

It is a pleasing solution to the problem at hand, but there is a fly in the ointment. This is an unproven theory: we have yet to see direct evidence of the majoron. Some indirect evidence may come into view at CERN's Large Hadron Collider, but our experiments are not yet powerful enough to recreate the conditions at the time before annihilation took place and observe the majoron at work. When it comes to working out why there is something rather than nothing, time travel back to the first moments of creation might provide our only dependable answer.

Do we live in a simulation?
Human nature, the laws of physics, and the march of technological progress

In 1998, almost no one you'd meet on the street would have given this question a moment's thought. By the end of 1999, the possibility had been discussed by millions of people around the globe. Why? Because they had seen *The Matrix*. The central premise of the film is that the human population of Earth is lying in vats of nutrient, their energy being harvested by a race of machines.

To keep us from reacting to this horror, we are granted an existence in a simulated reality, accessed via a direct connection to our brains. All our conscious experiences, then, are nothing more than the product of a computer program.

It's not an unprecedented idea. Philosophers since Descartes have argued about whether our perception of reality could be the product of deception, and science-fiction writers have used a similar premise many times. In 1966, for example, Philip K. Dick published a story where people bought 'implanted memories' that enabled them to experience things they had never done. The TV series *Doctor Who* introduced a massive computer system called the 'Matrix' in 1976; this could also be directly connected to the brain to allow out-of-body experiences.

But the 1999 movie *The Matrix* obviously hit the screens at just the right time. Within a few years of its release, physicists were discussing the idea at scientific conferences, and every time they did, it was the movie that was referenced. Strange as that may seem, there was good reason. The idea that we live in a simulated reality was one of the few plausible answers to a very old question that had just resurfaced in physics.

Looking out at the universe, astronomers have noticed something strange. They almost hesitate to mention it, but it is like an elephant in the room, and has to be acknowledged. This universe is remarkably good for us. Change it a little bit – tweak one of the laws of nature, say – and we simply wouldn't have arisen. It is almost as if the universe was purposely designed for our habitation. If that is the case, could the designer be a race of super-intelligent beings who have some reason – maybe work, maybe pleasure – to will our existence?

It's a big 'if', of course – perhaps the biggest 'if' in physics. The discussion of that 'if' even has a name: the 'anthropic principle'. It's a misnomer really. For starters, it's more of a suggestion than a principle. And, although anthropic means 'human-centred', that's not really what it's about. The person who coined the term, astrophysicist Brandon Carter, meant it to encompass not just human life, but the existence of intelligent life in general.

Carter came up with the anthropic principle at a time when physicists were coming to terms with a new paradigm: the Big Bang. Until the idea of a beginning to the universe was widely accepted, physicists had assumed there was no such thing as a 'special' time in the universe's history. The universe had always existed, and would always exist, pretty much as it is.

With the 1963 discovery of the cosmic microwave background radiation, though, everything changed. Once the radiation was recognized as an echo of the moment of creation, the universe was seen to have an unfolding history, punctuated by significant events. The trouble was, one of the central premises of astronomy has always been the Copernican principle, which asserts that humans do not hold any special place in space, nor in time. With the Big Bang, the Copernican principle was under threat.

A special universe?

But, Carter said, whatever our prejudices, we have to acknowledge there is something special about our relationship with the universe. 'Although our situation is not necessarily central, it is inevitably privileged to some extent,' he told an assembly of scientists in 1974. That privilege comes, first, through the laws that govern the universe's evolution.

There are a number of reasons why one might think that these laws were designed to give us a comfortable existence. The first is the rather convenient strength of gravity. After the Big Bang, space was expanding, forcing all the particles of matter further and further away from each other. The force of gravity was working against that expansion, however: the mutual gravitational attraction of the particles pulled them towards each other.

There are three ways in which this could have worked out. First, the expansion of space could have overwhelmed gravity's pull. In this scenario, known as the 'open' universe, every particle of matter would be pushed further and further apart, and the increasing separation would make the gravitational pull weaker and weaker. In this situation, galaxies – maybe even the stars themselves – would not have formed.

What if gravity's pull overwhelmed the push of expanding space? Then stars and galaxies might have briefly formed, but the strength of gravity means they would have quickly collapsed in on themselves and each other, and the universe would have imploded in a huge gravitational crunch. This is the 'closed' universe.

The third, 'critical' scenario involves a delicate balance between push and pull. Here the density of matter in the universe is such that, just after the Big Bang, the gravitational pull almost perfectly offsets the expansion of space. It pulls matter together just enough for stars to form, and for the stars to gather into galaxies. Thanks to their mutual gravitational pull, the expansion of the space between them is slowed, and the universe is granted a long and fruitful life.

A cosmic coincidence

So, what is the difference between these scenarios? When astronomers crunch the numbers, they first look at the critical universe. For this, they need to examine the density of matter in the universe, a parameter they call 'Omega'. It turns out that, for the critical universe scenario to occur, Omega had to have a particular value at one second after the Big Bang. Astronomers set this at one. And if Omega had been greater or less than one by an astonishingly small amount – one in a million billion – the universe would have either crunched closed or flung matter far apart way before life could establish itself in the benign environment surrounding a young star such as our sun.

It's not the only cosmic coincidence. If the strength of gravity is conveniently but finely balanced against the initial expansion of space, allowing stars like our sun to form, consider the efficiency with which the sun releases energy by fusing hydrogen atoms to form helium. The efficiency is around 0.007. That is, when the atomic masses of the hydrogen atoms are compared to the mass of the new-formed helium, 0.7 per cent has disappeared. This is the energy – mainly heat – that powers life on Earth.

So how much leeway is there here? Raising the efficiency

of transformation means allowing a slightly stronger 'glue' between the particles in the nucleus of an atom. If the efficiency were higher than 0.008, all the hydrogen created in the Big Bang would have been turned into helium almost immediately, and there would be none left to burn in stars. It would give a dead universe, in other words. Going the other way, lowering the efficiency to 0.006 would mean nuclear glue so weak that helium would never form, and the sun would never ignite. Again, no life would be possible.

Then there is the fact that the electric force is around 10^{40} times bigger than gravity. This gives atoms their fundamental characteristics. There is a mutual repulsion between the positively charged protons in the atomic nucleus. But there is also a mutual attraction due to gravity. Alter the ratio between them by a small amount, and you change the characteristics of atoms so much that it alters the characteristics of stars – go one way and you would create a universe where planets don't form around stars like our sun. Go in the other direction and you threaten the existence of the supernovae that forged the carbon atoms that underpin the chemistry of life. There are other examples. Reduce the neutron mass by a fraction of 1 per cent, and no atoms would form, for example.

Monkeying with the universe

It all sounds like a fix, doesn't it? The great English astronomer Fred Hoyle thought so. He once complained that the universe was so bio-friendly that it looked like 'a put-up job'. Someone or something, he suggested, was 'monkeying' with the laws of physics to facilitate the production of life.

So what does a scientist do about this? Besides saying God did it – which leads scientists nowhere in the quest for an answer – there are three options. The first option is to turn the problem on its head. We wouldn't be around to worry about these things if the universe were any different. Of course it is so precisely balanced for life. We could not be in a universe that was any different. Such an approach forces us to consider the existence of other universes where the laws of physics give different values to those crucial numbers. Besides being dead universes, however, they are also scientific dead ends. We cannot access them, so we have to content ourselves with not finding a satisfactory answer to the question of our universe's fine-tuning for life. A second approach is similarly unsatisfactory: we put the fine-tuning down to the existence of a supernatural designer, a being that transcends the natural laws. Here, too, we have no hope of discerning whether the approach is the right one.

The third option is the one we have been working towards: that the universe is so well suited to our existence because it was designed for our existence. The designers in this case are not deities. They are beings like us. Only much, much more advanced in their control of technology. So advanced, in fact, that they can create two amazing things. First, beings that exhibit what we consider to be consciousness. And second, a world for those beings to experience with their consciousness. This is the logical sequence known as the simulation argument. The first person to pull it all together was a philosopher called Nick Bostrom. In 2001, he began circulating a paper entitled 'Are You Living in a Computer Simulation?'. His answer was, yes, quite possibly.

Creating the world anew

Bostrom's argument is fairly straightforward. Stop and think about the computing power now at your disposal. Compare that to the power available a decade ago. What about two decades ago? Now translate that into the future. If our civilization survives the next millennium, the computing power available to its population will be of a magnitude that is unimaginable to us today.

Now come back to the present. What is one of the most popular types of computer games? Simulation. Take the extraordinary success of the simulation *Second Life*, for example. It gives people the opportunity for an alternative existence – an opportunity that millions have grabbed with both hands. Other simulation games allow you to play the deity, controlling others, or just watching how their fates unfold. Something about the human mind loves to get involved in another world. And why should things be any different one thousand years into the future?

Bostrom's argument is that one out of the following three

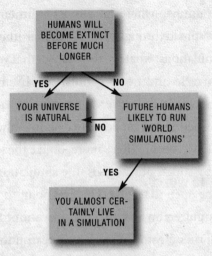

The simulation argument

propositions has to be true. The first is that humans are over-whelmingly likely to become extinct before reaching a level of sophistication where they are able to run computer sim-ulations – virtual reality – that would mirror what we experience as reality. The second is that any such civilizations that survive are extremely unlikely to run such simulations. The third is that we are almost certainly living in such a computer simulation.

The first proposition seems unlikely. There is no a priori reason why we will necessarily wipe ourselves out, or be wiped out. The second seems even more unlikely: our own delight in simulations gives no room for supposing that, given even more simulating power, we won't use it. Which leaves the third proposition. Given the fact that we are talking about a far future, where an almost infinite number of civilizations spread throughout the 'original' universe will be running simulations, what are the chances that we are in that original universe and not a simulation? Infinitesimal. In other words, we are almost certainly living in a simulation.

It's not something to get depressed about – the world is as real as it has ever been. What's more, unlike the ideas of universes run by supernatural gods, the simulation argument just might be open to testing. The first point to recognize is that it does indeed answer the question about fine-tuning. The simulation's creators must have a reason to create it. It seems sensible to suggest, therefore, that the overwhelming majority of simulations will have to work well enough to be interesting to their creators and users. Our experience with creating simulation environments suggests that this means populating them with beings that can enjoy their 'existence', which, in turn, tends to involve an ability to interact with the simulated world and its inhabitants.

A plausible simulation will therefore encourage the development of something we would regard as complex life. As we have seen with our look at the laws of nature, that gives a fairly narrow range of possibilities for the set-up. That at least provides a plausible explanation for the fine-tuning. Now we have to look for a scientific test for such an explanation. Again, this can be found within our own experience of creating simulations.

Conservative computing

One of the central rules of programming is that you don't waste precious computing resources. That means that any simulation will not be infinitely smooth. It will be built well enough to give its conscious avatars a sense of continuity in the world around them – but no better than is necessary. That means a sudden, close look might expose the gaps in the programming.

We may, in fact, have already done so. We already know that the theories we have devised to describe our reality have apparent inconsistencies. The quantum world, which seems to describe things we encounter at subatomic scales, for instance, does not make sense to the human mind. It allows particles to have multiple existences, occupying two spatial positions at the same time or simultaneously moving in opposite directions.

Similarly, relativity, which we use to describe reality when we are considering large, cosmological scales, fails to describe the most extreme of cosmological conditions, such as the interior of a black hole, or the geometry of the moment of the Big Bang. Could it be that these frustrating limitations to our theories reflect the limits of the programming behind our reality?

There is further evidence to consider. One of the most significant aims of modern science is to 'unify' the laws of physics. At the moment, the main thrust of that is to marry together relativity and quantum theory. However, it is a marriage that no one has yet managed to broker. Might that be because it is fundamentally impossible?

When creating today's simulations, programmers use a particular method for coding the finer details – the movement of hairs in a polar bear's fur, say. The methods for creating a facsimile of a pastoral landscape are different. Similarly, the creators of our simulation may have used different methods for programming our reality on different scales, so we should not expect to be able to marry them together. If that is the case, the frustrations of science might be a clue to the nature of our existence.

A further clue might be found in our genetic code. Our DNA tends to make mistakes when replicating. Left uncorrected, these mistakes would be enough to give any species a short shelf life – perhaps too short to evolve. The simulated story of life would have crashed quickly had it not been for error-correcting routines embedded in the function of our genes. We do the same with our computer programs: we incorporate error-correcting routines that put things right before things go irretrievably awry. It is not a big stretch, therefore, to imagine that the simulation's programmers would have to employ the same methods.

One suggestion that has been made by serious physicists is that a correction to the simulation might create cracks, or even breaks, in the laws of physics. Some things might not behave as expected. Have we made any such observations?

As a matter of fact, yes. Astronomers have suggested, for instance, that the light reaching Earth from the furthest galaxies observable shows signs that the laws of physics have suffered a tweak at some point in the distant past. The light was emitted 12 billion years ago, and its interactions with matter during its journey across the universe have a slightly different character than one might reasonably expect.

The observation seems to suggest that one of the constants of physics, the constant that governs the fine details of how light and matter interact, was subtly different in the past. Is this a programming error, or part of an error-correction routine? Though the scientific inference about the varying constant seems solid enough, the suggestion that it provides support for the idea that we live in a simulation remains controversial.

None of these 'tests' are knock-down convincing. The idea that we are living in a computer simulation is an intriguing one, and in many ways it offers a highly plausible answer to one of the most vexing problems of modern physics. Whether it can be proved or falsified remains an open question. Maybe that's why some philosophers have argued that the only way we will know for sure is if the humans propagating the idea are mysteriously 'deleted' from the simulation because they pose a threat to its continued success. Others have made a similarly playful, but far more appealing suggestion. Now that we have made this discovery, it seems entirely possible that we could soon find a huge message rending the sky asunder: 'Congratulations: please proceed to Level 2'.

Which is nature's strongest force?

The ties that bind the universe, and their origin in the superforce

It's a question straight out of Hollywood. Take two imposing but very different beasts, and set them against one another. We have had Alien vs Predator and King Kong vs Godzilla; how about Gravity vs the Strong Force? Or the Weak Nuclear Force vs Electromagnetism? You won't be surprised to hear that the answers to such questions are unattainable. The reason for that might come as a surprise, however.

If physicists' suspicions are proved right, we are not dealing with four forces, but one. Just as a skilled puppeteer can control more than one marionette, there seems to be one superforce behind what we see as the different forces of nature. It could be that gravity, electromagnetism and the strong and weak nuclear forces (see table: *How the Superforce Split*, p. 266) were once united.

In the preface to his great work *Philosophiae Naturalis Principia Mathematica*, Newton wrote that he harboured a deep suspicion that all the phenomena of nature 'depend upon certain forces by which the particles of bodies, by some causes hitherto unknown, are either mutually impelled towards each other, and cohere in regular figures, or are repelled and recede from each other.' The forces of nature, in other words, are at the core of physics.

This idea was in marked contrast to what had gone before. The Greek mode of scientific investigation was to assume and respect the role of a 'prime mover', an ultimate cause that also governed notions of justice and morality. Seeking purely physical mechanisms for natural events, without searching for the ethical and moral dimensions they related to, was just not done. But we now know that the forces of physics hold true for everyone but mean nothing in moral terms. Gravity, to paraphrase the Gospel of Matthew, 'sendeth rain on the just and on the unjust'.

Not all the forces are so inclusive. The electromagnetic force, for example, only acts between particles containing electrical charge. The strong force only acts over a short range, and between the particles in the nucleus. This raises a question. If they are all so different, why do we believe they are all of the same origin? To answer that, we look first at our ideas of gravity – and where they fall short.

The taming of gravity

Gravity, to us the weakest of the forces, was the first to be tamed. Newton made the initial move in his universal law of gravitation, offering a formula that described how any bodies with mass would interact. In Newton's scheme, the pull of gravity accounted for the motions of the planets with an astonishing degree of accuracy. Newton's gravitational ideas fell short in two ways, however. One was that they offered a description but no *explanation* of gravity. The other was that they did not describe every facet of how gravity works in the universe: some phenomena defied explanation.

The precession of the perihelion of Mercury is perhaps the most famous example. The perihelion is the point of closest approach in an elliptical orbit. Mercury's trip round the sun has just such a point, which moves, or precesses, with successive orbits. The precession is a result of the gravitational pull of the other planets in the solar system, and, in 1845, the French astronomer Urbain Joseph Le Verrier used Newton's law to work out what it should be. There seemed to be an error. Le Verrier's calculation missed the observed precession by 43 seconds of an arc per century. Every hundred years, the calculations were out by just one hundredth of a degree, but they were wrong nonetheless.

Fortunately, Einstein's general theory of relativity provided the required correction. Relativity describes the gravitational fields as arising from the influence of mass and energy on the fabric of the universe: gravity comes from a warping of space–time. It is an astonishingly successful theory, and has never failed an experimental test. Nonetheless, for all relativity's grand successes in describing what we see in the universe, a proper explanation for the why and how of gravity remains elusive. And until we have one, we cannot be sure that gravity really is so weak – especially when we examine the next force to succumb to science.

Charged and ready

Electromagnetism is a much stronger force than gravity. Take two electrons: the electromagnetic repulsion between them is 10^{43} times larger than their mutual gravitational attraction. But this relative strength may be an illusion. The clue lies in the fact that electromagnetism is a unification of two theories: electricity and magnetism.

In the 1840s, the English physicist Michael Faraday had come up with the concept of a field to explain why iron filings formed lines when scattered around a magnet. To Faraday, these 'lines of force' were associated with some physical properties of the space around the magnet. The link to electricity came easily: Faraday also discovered that a changing magnetic field creates an electric field.

But there was a complication. When Faraday's friend James Clerk Maxwell tried to pull Faraday's discoveries, and the equations that described them, together, he could only make sense of the result if he added another factor into the mix. It is not enough that changing magnetic fields create electric fields. The converse must also be true: changing electric fields, Maxwell said, must create magnetic fields.

Maxwell's new equations glowed with a beautiful consistency: electricity and magnetism were two sides of the same coin. This unification led to another beautiful result. When Maxwell looked at the consequence of a changing magnetic field growing an electric field, which grew a magnetic field in turn, and so on for infinity, he realized he had discovered the root of electromagnetic radiation. What's more, the speed of propagation of this disturbance was the speed of light. Light, it became immediately clear, is an electromagnetic wave.

The significance of this discovery is hard to overestimate. It led to the discovery of the electromagnetic spectrum, to radio waves and gamma rays and everything in-between. It showed how energy could be transferred from point to point through space, doing away with the idea of some ghostly interactions that had no physical source. Perhaps most importantly, it paved the way for an instant revolution in physics. Maxwell's equations didn't work when the source of radiation was moving relative to an observer, an observation that prompted Einstein to resolve the anomaly with special relativity (see *What is Time?*) in 1905. What's more, the unification of electricity and magnetism was only the start. We now know that another of nature's forces is delivered from the same hand.

Einstein's Achilles heel

Einstein was highly motivated by the idea of unification. After the success of relativity, he spent his life trying to construct a 'unified field theory' that pulled electromagnetism out of the geometry of space–time, just as he had done with gravity. As a result, he and his few followers ignored the development of quantum theory. Einstein had never liked it, and hoped it would go away.

But it didn't, and explorations of the new theory, along with the fast post-war development of particle physics, pointed to the existence of two new forces: the strong and weak nuclear forces. Einstein never addressed these, but carried on playing exclusively with electromagnetism and gravity. By the time he died in 1955, physics had moved on without him.

It is a particular shame because the weak nuclear force, which acts between the particles of the nucleus – the neutron and the proton – and has an extremely short range of 10^{-17} metres, is now known to be closely related to the electromagnetic force. We know this because the weak force is responsible for 'beta' radiation, where an atom emits an electron or its positively charged counterpart, a positron. The beta-emission of an electron involves a neutron turning into a proton, which can only happen if a 'W boson', the source of the weak force, is emitted first: it is this particle that then decays to produce the electron.

The link was made stronger when we realized that the weak force and the electromagnetic force result from the same process, known to physicists as 'spontaneous symmetry-breaking'. This is rather like what happens when you assemble a crowd of strangers in a room. As they get talking, some will find common points of interest in one area, some in another, and, given enough time, they will form into distinct groups that end up talking about different things. Initially, there was 'symmetry': there was nothing to distinguish the strangers from one another, no way to group them. But, as they talked, that symmetry was spontaneously broken, and groups formed.

In the 1960s, Steven Weinberg, Sheldon Glashow and Abdus Salaam showed that the same process of spontaneous symmetry breaking created the electromagnetic and weak forces from another force. They named it the 'electroweak force', and suggested that it had only existed in its unbroken form in the high-energy conditions at the beginning of the universe. The work was a masterstroke, and won the trio the 1979 Nobel Prize in physics. The theory made specific theoretical predictions: the existence of the W and Z bosons, for example, which were found, complete with all the assumed characteristics, in 1983.

Perhaps most important of all, though, this breakthrough suggested that the seemingly different forces might not be all that different at heart, even though the weak force acts over the shortest ranges, and on uncharged neutrons, whereas the electromagnetic force acts over enormous distances, and on charged particles. In fact, not only can we not say which is the stronger force, we suddenly find ourselves facing a shocking question.

If the electromagnetic and the weak forces were once the same force, who is to say that spontaneous symmetry breaking didn't give rise to all of the forces of nature? Perhaps we can't say one force is the strongest, simply because they are all manifestations of one ancient superforce. In order to explore that possibility, we have to consider the remaining element: the strong nuclear force.

The nuclear bind

Just as the weak force had to exist in order to explain beta-decay, the mutual repulsion between protons in a nucleus made the strong force a necessity, otherwise the nucleus could not hold together. Strong is an appropriate name for the force: typically, it appears to have a hundred times the strength of the electromagnetic force that would tear the nucleus apart. Measuring its strength was the easy part of taming the strong force, however; explaining its existence was much more difficult. It's not enough to know that such a gargantuan force is the only way that a nucleus can hold together. What creates it?

The ideas behind this strong force were developed in the early 1970s. It was known that quarks make up the protons and neutrons in the nucleus. Each quark has a characteristic that physicists call its colour. For this reason, the theory tying the strong force to the quarks is called 'quantum chromo-dynamics', or QCD. According to QCD, the strong force binds quarks together using an interaction that, unlike the electromagnetic and gravitational forces, does not diminish with distance. The force grows stronger as quarks move further apart just as if they were connected by a spring.

This peculiar property, which emerges from the equations of QCD, gives the strong force the power to bind quarks together wherever they might be found. Its nature is borne out by the fact that, despite many searches, we have never found a free-roaming lone quark. QCD says that the strong force is created by a boson known as a gluon. Gluons were seen in experiments for the first time in 1979. The theory was already on a firm footing by that time, however: when quarks, complete with their predicted characteristics, began to be spotted in particle accelerators in the late 1960s and early 1970s, QCD was considered a proven theory.

But what really excited physicists was the fact that QCD is built on the same symmetry-breaking idea as the electroweak force. It seemed entirely plausible that they could be closely related – and pulled together into one description of the behaviour of matter: the 'grand unified theory', or GUT. And here is where the quest hits the skids. After three decades of searching, we are still not sure if the strong force really is from the same stable as the electroweak force.

The struggle towards unification

The problem is that unification is far from straightforward. What is required is another symmetry, like the strangers gathered in a room scenario – but this time there are even more of them. Somehow this bunch of indistinguishable strangers has to spontaneously break up in a way that describes five different types of particles – the three differently coloured quarks and the electron and its associated neutrino – and three forces.

The unification is almost impossible to recreate on Earth: reaching the energy for this symmetry breaking requires particle accelerators 100 billion times more powerful than the Large Hadron Collider (LHC), our most powerful atom-smasher. However, there are other ways to test the idea. According to any grand unified theory, quarks must be able to change into electrons and neutrinos, and physicists' best-looking candidate for this grand theory (known as SU(5) because of the five particles that arise from it) has just such a process up its sleeve. It involves the proton in a kind of radioactive decay, and makes a prediction about how often that will happen.

It's just a shame it gets it so very wrong. The theory says a proton will last approximately 10^{33} years before decay. About a quarter of a century ago, physicists built huge tanks of highly purified water surrounded by detectors that would register such an event occurring. From the theory, and the number of protons in their tanks, they expected a few decays per year. So far, though, they have seen nothing. We still have another shot, though – and this one might be vindicated in the LHC. It is called 'supersymmetry'.

SUSY comes calling

Supersymmetry arises from the fact that physicists split particles into two camps: the fermions, such as the electron and the quarks, which make up matter; and the bosons, such as the photon and the gluon, which create the forces. These two different kinds of particles follow two different sets of rules. And supersymmetry, or SUSY, says each one has a 'superpartner' from the other camp that will behave the same in any experiment.

That is possible because the essential difference between fermions and bosons comes from the quantum property known as spin. Bosons have integer spin – 1, 2, 3 and so on – while fermions have spins that are half-integers: 1/2, 3/2 etc. SUSY involves applying a kind of perspective change, something akin to looking at a clock from the front or the back. This change of view alters quantum spin (just as the sense of rotation of the clock's hands is different when viewed from the back), but not other qualities, such as electric charge or quark colour.

It might sound like a convenient fiction, but it is a highly respected mode of thinking that stands amongst the best

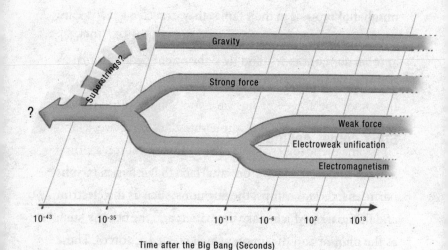

Time after the Big Bang (Seconds)

How the superforce split

ideas in physics. The burning question, of course, is whether it is true. Besides spin, one other property of the superpartner particles is changed: their mass. They are much, much heavier than the set of particles that we are familiar with. That means that, thanks to $E = mc^2$, they will only exist at high energies. Thankfully, however, the LHC's collision energy of 14 TeV should be high enough to see the lightest of them, which are thought to come into play at around 1 TeV.

Though it sounds promising, these particles are still difficult to detect. They hardly interact with normal matter, and will fly out of the machine almost without trace. That means the only hint of supersymmetry might be some energy missing from the LHC's detectors. Since other theories suggest that some normal particles might be disappearing into other, 'hidden' dimensions of reality, it's a recipe for false positives and missed sightings.

If we do see unquestionable sightings of supersymmetric particles, though, we can feel sure that the grand unified theory of the forces of nature is on solid ground. It will be entirely reasonable to assume that the strong, weak, and electromagnetic forces arise from a common source, a 'prime mover', as the Greeks would say. There is just one fly in the ointment, though. What about gravity? Is that also part of the unification, or is it a separate entity? If we can't say there is a strongest force, can we at least say that gravity is the weakest?

Gravity certainly is weak. When we draw the likely unification diagram for the forces, showing the energy where they (might) unite, it's hard to put gravity on there, unless your graph is bigger than the known universe. While the other forces converge from a factor of 100 or so apart, gravity is simply off the scale. But there is a get-out. It is enormously technical, but, boiling it down, it says that the gravitational interaction depends on mass, which is proportional to the energy involved. In the SUSY picture, when considering the high-energy conditions of unification, gravity drops into the picture at an alluring scale – almost, but not quite, where the other forces unite.

It's not a fully convincing answer, but it does suggest that gravity and all the other forces of nature might spring from one ultimate force. This superforce only existed in the first moments after the universe was formed. In that situation, asking which of the forces is stronger is like asking which of the particles is more particle-like. Though different, they are all aspects of one characteristic. Gravity vs electromagnetism just won't work; they appear to be fighting from the same corner.

What is the true nature of reality?

Beyond the quantum world lies the realm of information

This will be the last question physics answers – if, that is, it is even possible to do so. Constructing a grand unified theory of forces is all very well, and the hunt for quantum gravity, the theory that unites the science of the very small with the science of the very large, is exciting and useful. But neither of these will answer the fundamental question: what is reality made from?

Some may argue that this quest lies beyond the reaches of science. But it is in the very nature of physics to find answers to seemingly impossible questions. The history of physics is littered with 'impossible' tasks that have turned out to be very possible indeed. It is easy to forget that Archimedes stunned the ancient world with his innovative thinking. Whether or not he really worked out a way to tell if the king's crown was made from pure gold, he achieved enough of a scientific reputation to have his life protected by Roman mandate. Similarly, Newton's description of how gravity works seems obvious now, but its formulation was truly a *tour de force* at the time.

St Augustine described magnetism as not unlike a miracle, but now we know the microscopic processes behind the whole gamut of electromagnetic phenomena. The physics done in the past seems a little prosaic now; indeed, the concepts are so straightforward to us that we learn many of them as children. So, will schoolchildren of the future be bored in lessons on the fundamental nature of reality?

A perfect realm

The nature of reality is an avenue that humans have tried to explore since at least the time of the ancient Greeks, and it is to then that we can trace our own quest for the nature of reality. The Greeks had several schools of thought on this question. Perhaps most influential was Plato, who believed in a realm of perfect abstractions of physical entities. Everything in the material world drew its existence from these 'ideal forms', and everything was but a shadow of the ideal object.

This realm, accessible only through the training of the mind, was not just about physical objects, such as trees or mountains. It also applied to mathematical ideas: Plato envisioned an ideal mathematical reality populated by the ideal solids. These five geometrical shapes created a connection between the mathematical and the physical. In his dialogue *Timaeus*, for instance, Plato linked the cube to the Earth: 'To Earth, then, let us assign the cubic form, for

Earth is the most immovable of the four and the most plastic of all bodies, and that which has the most stable bases must of necessity be of such a nature.' A similar logic linked the tetrahedron to fire, and the icosahedron to water, the octahedron to air and the dodecahedron to a mysterious 'fifth element', or quintessence.

Though it may sound like mystical mumbo-jumbo, we cannot entirely dismiss the notion of realms of ideals. All the tools of modern physics that have been gathered for use in the quest for the ultimate nature of reality have their root in mathematics – and mathematicians still can't agree on whether mathematics is an invention of our minds or an abstract world into which mathematicians venture in order to make discoveries.

The mathematician Roger Penrose has suggested that, if we are to understand what reality is, we may need to address this fundamental issue. Our best description of reality might have to involve a kind of trinity, he suggests: physical reality is only discernible because of the 'mental reality' – or consciousness – constructed by our brains, and can only be described if we believe that our equations and laws of physics come from some 'mathematical reality' that exists in parallel with our physical world.

In a philosophical version of the paper, scissors, stone game, Penrose suggests there is a cyclic dependency between the three realities. Only through the equations of mathematics can we describe the fundamental physical particles

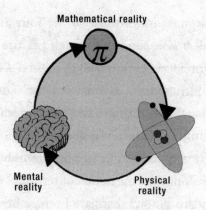

The trinity of reality

such as the electron, so mathematical reality trumps physical reality. But physical reality, in the form of the brain's neurons, gives rise to mental reality. And because mathematics is abstract, mental reality gives rise to mathematical reality. Mathematical trumps physical, physical trumps mental, and mental trumps mathematical.

A new reality

There is something, however, that seems to lie beyond all three of these notions of reality – something that pushes our notion of the ultimate nature of reality into an even more abstract realm. That something is information. We hold information in our minds, it can be manipulated mathematically and it is always wedded to physical things: information cannot exist without something – a splash of ink on paper,

DNA, a photon of light – to sit on. That's why, in 1991, the IBM researcher Rolf Landauer made an odd statement that still sounds odd today: 'Information is physical.'

What Landauer meant was that information is not some abstract concept, a convenient shorthand for what gets transferred in communications. Wherever you find information, it is inextricably linked to some physical system. Information is carried in the arrangement of molecules on a strand of DNA, enabling the propagation and evolution of life. It is encoded in the charge on a capacitor in an electrical circuit – allowing us to build the information storage and processing facilities we call computers. It is written into the quantum state of a photon of light, allowing telephone conversations to be sent through optical fibres. Wherever information exists, it takes a physical form.

The idea has come to be known as 'Landauer's principle', and it has sparked a revolutionary way of thinking about information. If information is physical, could it be that everything physical is actually just information? There are at least three good reasons to believe this to be true. First is the fact that information seems to be spookily connected to laws that govern the universe.

The speed of information

Perhaps the deepest notion in our understanding of the cosmos is special relativity's assertion that there is an unbreakable speed limit: the speed of light (see *What is Time?*). This has enabled us to make sense of countless phenomena in astronomy and cosmology. But it may just be that the limited speed of light is a result of the limited speed of information. Is the theory of relativity actually an offshoot of information theory?

Information theory didn't start out as something profound. It was developed by Claude Shannon, a mathematician and engineer who worked at Bell Labs in the 1940s. The main thrust of his work was to find ways to increase the speed at which information could be squeezed down a telephone wire, or through an electrical circuit. He developed techniques for 'compressing' information to optimize this, but also found fundamental limitations. Shannon discovered that each communications channel has a maximum capacity, and there is also a maximum efficiency with which information can be sent without it getting lost in transmission.

The measure of information is the 'bit', which is short for binary digit. Computers, for example, run on the binary number system: every number, every instruction, is encoded as a series of 0s and 1s. Though information can be stored in ways that offer more than two alternatives – DNA uses the four 'base' molecules, adenine, thymine, cytosine and guanine, for example – these can always be built from a binary system. The two-alternative system of the bit is the simplest, most fundamental means of storing and transmitting information.

The other important factor in information theory is the

'bandwidth' of the information channel. Whether it is in an Internet connection, or the connection between the memory and the processor in your computer, the bandwidth provides a measure of how many bits can pass through each second. Any channel for information will contain a certain amount of noise that will cause errors in the transmission of information. When NASA send radio signals through the Earth's atmosphere, for instance, the signal can be distorted by atmospheric conditions, turning a 0 into a 1 or vice-versa.

Shannon worked out that, given a particular signal-to-noise ratio and bandwidth, there is an upper limit on how fast information can be transmitted through the channel without any loss. The latest mobile phone and satellite TV systems work to within 1 per cent of this 'Shannon limit'. However, they cannot get to it, or get past it. It is a little like the speed of light in relativity: the closer you come to this fundamental limit, the harder it is to do any better.

Why should information be like the speed of light? Is it because, like light, information is related to the fundamental underlying structure of physical reality? That is certainly what a growing number of researchers believe – especially those who research black holes.

Where did the information go?

Black holes are the second reason to think information is part of the answer to our question about the nature of reality.

Nothing that falls within the spherical region known as the 'event horizon' surrounding a black hole ever escapes. That means black holes are, effectively, sinks for information. Everything they swallow has information encoded on it in the form of atomic states, spins of particles and so on. So what happens to that information?

In the 1970s, Stephen Hawking showed that black holes slowly evaporate, emitting 'Hawking radiation'. The trouble is, this radiation does not contain any information. The laws of physics dictate that information, like energy, cannot be destroyed, which means it must go somewhere. After decades of debate, physicists now believe that the information is encoded in the microscopic structure of space and time at the 'event horizon' of the black hole, the point of no return for in-falling matter or light.

Since the event horizon is a 2D structure – the surface of a sphere that surrounds the black hole – that means the information describing 3D objects such as atoms can be encoded on a 2D surface. Extrapolating from this idea, some researchers have demonstrated that the whole universe can be viewed in the same way. The boundary of our universe is essentially the 2D surface of a sphere. The information that appears to exist within the sphere could actually be held on the surrounding 2D surface. Just as an apparently 3D hologram results from a carefully designed projection of light onto a 2D surface, our 3D reality could well be a hologram projected from information held at the edge of the universe.

In other words, everything you think of as physical stems from information.

There is even a hint of experimental support for this idea. In 2008, US particle physicist Craig Hogan was trying to work out how we might test the holographic projection idea. He worked out that the boundary of the universe could hold only a limited amount of information, and that, when the information was projected into the 3D space of the universe, this limit would manifest as a kind of pixellation effect in our physical reality. We would, effectively, see the dots; space and time, Hogan suggested, should look grainy if you could see it on small enough scales.

The kinds of scales involved mean it would only be detectable in the most sensitive instruments we have – the gravitational wave detectors looking for the ripples in space and time that result from violent cosmological events such as the collision of two black holes. And so Hogan sent his best idea of how the graininess of space–time would affect these instruments to the scientists at GEO600, a gravitational wave detector sited in Hanover, Germany.

As it turned out, the GEO600 researchers had been having problems with noise in their detectors. And that noise had exactly the same characteristics as Hogan's anticipated signal. It's not yet conclusive evidence, but it suggests that the 'holographic principle' – that everything is ultimately composed of information that resides at the edge of the universe – is at least worth taking seriously.

Quantum information

The third reason to consider information so important comes from quantum theory, our best set of rules for how things behave on subatomic scales. Quantum theory has been astonishingly successful, its predictions matching experiments without fail. But it is not the final answer to understanding the nature of reality. Though it provides a way to describe what happens in subatomic systems, it does not tell us why things behave as they do (see *What Happened to Schrödinger's Cat?*). In fact, it leaves us perplexed about many aspects of the behaviour of these systems, giving room for philosophers to wax lyrical about the absence of objective reality, and the limits of experimental science.

There are more than half a dozen philosophical interpretations of the limited view that quantum theory gives us. There is no way to choose between them because all of them are consistent with all the experiments. The only way out, it seems, is to find what lies beneath quantum theory – and that appears to be information. There is an obvious link between quantum theory and information: with bits and quanta, both information theory and quantum theory rely on a fundamental, indivisible quantity. But there is also a more subtle connection. The strangeness of the quantum world could arise from limits on the amount of information carried by a quantum particle.

One reason for thinking this is the Heisenberg uncertainty

principle, which says that, if you know some things about a quantum system with perfect accuracy, there are other things that you cannot know at all (see *Is Everything Ultimately Random?*). Heisenberg deduced his principle from the equations of quantum theory, and we have so far had to accept that this is 'just the way it is'. By considering aspects of information theory, however, we can achieve a somewhat more satisfactory explanation.

A quantum particle such as an electron has a property called spin, which is binary (up or down) and can be measured in any of the three spatial dimensions. If an electron's spin can only carry one bit of information, the first measurement on the electron will use that bit up; there is no more spin information available to measurements in the other dimensions. The outcome of any such subsequent measurements will be random – exactly what is predicted by the Heisenberg uncertainty principle.

There are indications that information theory might also be able to make some sense of the puzzling phenomenon of quantum entanglement, which allows a 'spooky' link between two particles. Entanglement is certainly all about carrying and sharing information. Put simply, it dictates that, after two particles interact, each particle's quantum state – the full description of its position, momentum, spin and so on – resides not within that particle, but is shared between the two.

The spookiness of entanglement lies in the fact that the particles can be placed in a quantum state that is 'indefinite'.

Just as Schrödinger's cat is alive *and* dead until it is observed (see *What Happened to Schrödinger's Cat?*), the entangled pair can have a mixture of spins – they can be 'up' and 'down' at the same time – until someone measures the spin.

When a measurement forces one particle to a particular spin, the spin of the other particle will be made definite. Einstein hated this because it looks like the observation of one particle can change the state of another one, no matter how far apart they are (see *Can I Change the Universe with a Single Glance?*). If an entangled pair of particles can carry only a limited amount of information in their spin states, however, that provides a way out of the weirdness.

The quantum version of information theory says an entangled pair can carry only two bits of information. If those two bits encode something like 'the spins are the same when measured in the X dimension', and 'the spins are opposite when measured in the Y dimension', that gives a description of the spin states of both particles – but leaves no room for information about the spin of an individual particle.

That's why the first measurement appears to give a random result, yet the result of the second measurement can be predicted with perfect accuracy. Though it gives the illusion of a 'spooky' transfer of information between the particles, it's actually just that the first measurement gives us more information. Given the first result, and the nature of the link between the spins, the second particle's spin can be deduced with simple logic.

Quantum researchers have only just begun to appreciate that information might be the key to understanding their discipline, and don't yet have many solid explanations for how this might work. But if information does lie at the root of quantum theory, that seems somehow appropriate. We are living in what has been dubbed the 'information age', where optical fibres and satellite transmissions fire information around the world at astonishing speeds and intensities. All these technologies work because of our understanding of the quantum world – the laser and the microchip are both spin-offs from quantum theory. It seems only right that the last question in physics should tie information theory and quantum theory together.

Sophisticated sceptics

So where does this leave us in the search for the ultimate nature of reality? If you can describe anything as an 'it', as a real entity, it ultimately appears to come from a bit of information – or a large collection of them. We get, as the physicist John Archibald Wheeler put it, 'it from bit'. In 1990, Wheeler declared that, 'Tomorrow we will have learned to understand and express all of physics in the language of information.' That 'tomorrow' has not yet come, but perhaps it is appearing on the horizon at last.

However, we simply cannot know how far we are on the path to discovering the ultimate nature of reality. During this century our investigations of reality have taken us from the realm of the atomic to the subatomic, right down to the idea of energetic fluctuations in the fabric of space and time. It seems that the fundamental nature of reality goes deeper than this, into abstract notions of mathematics and information. But is that the end?

Physicists are painfully aware that any and all their conjectures could be a million miles from the truth. They work within the current limits of knowledge and the limits of the human imagination. Both seem to recede as we discover more about the world, but never disappear. If the end of physics is on the horizon today, it is worth remembering that it has always seemed to be there. It would be hubris to think we are taking the final steps towards understanding the very core of reality; there is undoubtedly plenty of distance left for physicists to cover. But when the journey is so deeply fascinating, that can only be cause for celebration.

Glossary

absolute zero −273 celsius: the temperature at which no substance contains any heat energy.

alpha radiation A non-penetrating but potentially damaging form of radiation. Alpha particles are identical to the helium nucleus, containing two protons and two neutrons.

anthropic principle The notion that asking why the universe is as we see it is essentially pointless because we could not exist in any universe that is substantially different.

antimatter The 'nemesis' of matter: every matter particle has an antimatter counterpart. When a particle and its antiparticle collide, they annihilate one another.

beta radiation Radiation composed of high energy electrons or their antiparticles (positrons) that is readily stopped by a thin sheet of metal. It is caused by nuclear processes that involve the weak nuclear force.

Big Bang The moment, nearly 14 billion years ago, when the universe came into existence.

black hole A patch of space–time where gravity is so strong that nothing (not even light) can escape. Black holes are often formed when a giant star collapses under the pull of its own gravity.

boson A particle with a quantum 'spin' number that is an integer. Bosons are the particles through which forces act.

closed timelike curve A region of space–time that, if followed, brings one back to the same moment in time.

compactification The process by which the 'extra' dimensions suggested by many modern theories of physics remain undetected.

cosmic microwave background radiation The radiation released around 300,000 years after the Big Bang. The CMB radiation contains many clues to the nature of the early universe.

dark energy A mysterious form of energy that many physicists believe to be responsible for the accelerating expansion of the universe.

dark matter Hypothetical matter that, according to most astronomers, makes up nearly a quarter of the mass/energy in the universe. Dark matter is scattered throughout the universe, but concentrated in haloes around galaxies and galaxy clusters.

double slit experiment An experiment originally created to demonstrate the wave nature of light. It has also been used to show that quantum particles have wave-like characteristics.

electromagnetic force A force that causes particles with an electric charge to repel or attract one another, depending on the signs of their charge.

electron A subatomic particle, believed to carry the fundamental unit of electric charge.

electroweak interaction A force, thought to exist in the hot conditions of the early universe, which gave rise to the electromagnetic and weak forces.

entanglement A phenomenon that occurs when two quantum

particles interact. They become linked, with information about both particles residing in each one.

entropy A measure of the disorder of a physical system. The entropy of a closed system always increases.

ether A fluid that was thought to fill the universe and carry electromagnetic waves such as light. Its existence was disproved in 1887.

fermion A matter particle whose quantum spin is half of a whole number.

fractal A geometry that looks the same whatever the scale on which it is viewed.

fundamental forces The four fundamental forces are: strong; electromagnetic; weak; gravity.

gamma radiation A highly penetrating radiation that results from the emission of high-energy (gamma ray) photons in nuclear reactions.

gas A fluid composed of particles that are only very weakly attracted to one another.

general relativity Albert Einstein's description of the warping of space and time by the presence of mass and energy.

geodynamo The turbulent ball of churning molten iron that occupies the inner core of the Earth and creates a self-sustaining magnetic field.

gluon The particle that mediates the strong nuclear force.

grand unified theories (GUTs) Theories that attempt to define how three of the forces of nature (strong, weak and electro-magnetic) were once one in the conditions of the early universe.

gravity The attractive force that acts between particles with mass or energy.

Heisenberg uncertainty principle A rule that imposes a limit on the accuracy with which certain combinations of properties can be known for any quantum particle or system.

hidden extra dimensions Certain theories suggest that there are more spatial dimensions than the ones (up–down; back–forth; across) we experience.

Higgs boson A particle that is thought to mediate the Higgs field, the source of certain types of mass.

inflation A theory that suggests the universe went through a period of ultra-fast expansion just after the Big Bang.

interference A wave phenomenon where two waves interact, producing a wave whose properties depend on the size and relative phase of the original waves.

inverse square law When a force between two spatially separated objects diminishes, by an amount proportional to the square of the distance between them, as they are moved further apart.

kinetic energy Energy associated with motion.

liquid A state of matter in which particles have some attractive forces acting between them, and can only be further separated by an input of energy.

magnetosphere The magnetic field surrounding the Earth.

many worlds theory The idea, originated by Hugh Everett III, that every quantum event creates a new and separate universe. Only quantum particles such as electrons are able to sense the existence of these other worlds.

mass The attribute of an object that gives resistance to acceleration (inertial mass) or responds to and creates the gravitational force (gravitational mass). Physicists believe the two types of mass are equivalent but this is not proven.

Maxwell's equations The equations, laid out by James Clerk

Maxwell, that define how electric and magnetic fields behave and interact.

momentum The product of the mass and velocity of a particle.

multiverse A universe composed of myriad smaller universes, usually without any connection that allows passage between them.

neutron A particle composed of three quarks, but with no net electric charge.

nucleus The central core of an atom where almost all of its mass is concentrated.

particle accelerator A machine used to smash subatomic particles into one another. Analysis of the resulting debris can give clues to the fundamental constituents of matter and the nature of the universe.

perpetual motion machine A hypothetical machine that would do useful work with no energy being supplied.

photon A particle of light or other electromagnetic energy.

potential energy The energy an object contains because of its position within a field (usually gravitational or electric).

proton A positively charged particle composed of three quarks.

quantum A fundamental unit, originally of energy but now applied to anything indivisible, such as the electron charge, in subatomic physics.

quantum chromodynamics (QCD) A theory that describes the interactions of quarks and gluons in nuclear physics.

quantum electrodynamics A theory that describes the interaction of the electromagnetic force with matter.

quantum gravity A theory that will bring together quantum theory and relativity to create an overarching description of how things behave over very small and very large scales. String theory is one attempt to build a quantum gravity theory.

quark A subatomic particle associated with atomic nuclei. Protons and neutrons are each composed of three quarks.

Schrödinger equation The equation used to define how matter behaves at subatomic scales.

solar wind A stream of charged particles that fly off from the surface of the sun. The solar wind is responsible for the *aurora borealis* (the northern and southern lights).

solid The state of matter where particles are strongly bound together in a rigid structure.

space–time The four dimensional fabric of the universe. Any one point in space–time is known as an 'event': a specific time and location.

special relativity Einstein's 1905 theory that laid out a way for the laws of physics to be the same for all observers, regardless of how they were moving through space and time. He later generalized the theory to include the effects of gravitational fields.

standard model The theory of physics that describes the properties of all known particles, and their interactions.

string theory A description of the subatomic world where particles and forces are described by the vibrations of strings and loops of energy.

strong force The short-ranged force that binds quarks into protons and neutrons, and overcomes the repulsion of the positively charged protons to hold atomic nuclei together.

superconductor A material that offers no resistance to the flow of electrical current.

superfluid A fluid that exhibits no friction. The motion of a stirred superfluid continues indefinitely.

supersymmetry A theory that gives known particles a hypothetical 'superpartner' that is much heavier but has a related quantum spin. Supersymmetry is an important part of the attempts to unify all the forces of nature.

thermodynamics The study of how energy will create heat, work and entropy in a physical system.

time dilation A prediction of special relativity, borne out in experiments, that says the flow of time is altered by relative motion or by the presence of a gravitational field.

vacuum energy The energy inherent in otherwise empty space–time. It is due to the Heisenberg uncertainty principle, which does not allow space–time to have zero energy for any finite period of time.

weak force The short-range force that affects all matter particles. It is responsible for, among other things, some forms of radiation.

wormhole A hypothetical shortcut through space–time that could be used for time travel.

Index